TALK TO YOUR Cat

How to communicate with your pet

SUSIE GREEN

CICO BOOKS
London

5070569

First published in Great Britain in 2005
by Cico Books
32 Great Sutton Street
London EC1V 0NB
© 2005 Cico Books
Text © 2005 Susie Green

10 9 8 7 6 5 4 3 2 1
ISBN 1-904991-16-5

CIP data for this book is available from
the British Library

Edited by Robin Gurdon
Artworks by Trina Dalziel and Philip Hood
Photographs reproduced by kind
permission of Creative Image Library,
pp. 6, 46, 88, 112, 132, 136; Thompson
Animal Photography, pp. 28, 70;
Zoological Society of London, p. 13.
Cover photographs © Creative
Image Library
Cover design by Jerry Goldie
Graphic Design
Design by Ian Midson

Printed and bound in Singapore

Spike

Dedicated to:

Preservers, protectors, and
conservationists of felines big
and small, wild and free, exotic
and commonplace, tame and
savage, and to the ever wild-at-
heart parodoxical Felis catus.

"The greatness of a nation and its
moral progress can be judged by the
way its animals are treated."
Mahatma Gandhi (1869–1948)

CONTENTS

CHAPTER ONE

THE EVOLUTION OF THE CAT

"Cats are intended to teach us
that not everything in nature has
a function."
—Garrison Keillor (1942–)

The origins of what is called the domestic cat are, as is fitting to so esoteric a creature, shrouded in mystery-to this day authorities disagree on her exact classification, and her true ancestry.

Around 65 million years ago, the dinosaurs, for reasons still surrounded by dissent, became extinct and warm-blooded creatures began their long march to world dominance. Ten million years passed before these first mammals began to specialize into different ecological niches. At this point a split occurred, one branch containing what we think of as more catlike creatures, the other those who were more bear- and doglike. It was to be around another 20 million years before *Nimravidae* (now extinct), and the fearsome *Felidae*, direct ancestors of the domestic cat, finally evolved.

Showing the development of the felidae direct ancestor through millions of years, this is Alan Turner's hypothesis from The Big Cats and their Fossil Relatives.

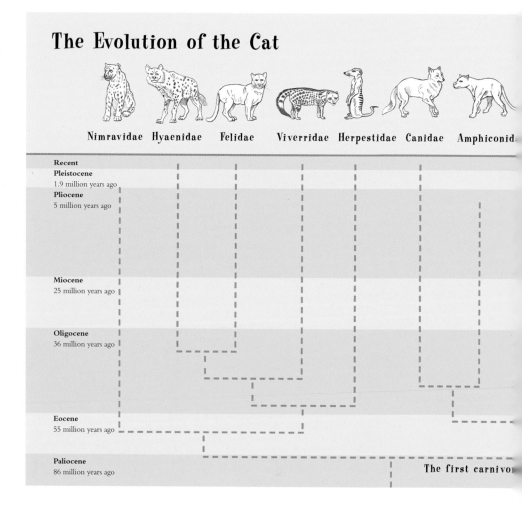

The Evolution of the Cat

Nimravidae Hyaenidae Felidae Viverridae Herpestidae Canidae Amphiconid.

Recent

Pleistocene
1.9 million years ago

Pliocene
5 million years ago

Miocene
25 million years ago

Oligocene
36 million years ago

Eocene
55 million years ago

Paliocene
86 million years ago

The first carnivo

How the Cat Family Developed

The *Felidae* family itself split into two subfamilies. The *Machairodontinae*, or sabertoothed cats, had enormous, flat, sheared, and elongated canines, ideal for slicing flesh but seemingly far too fragile to seize prey—some researchers believe these cats had to rush at their prey and knock them over before they were able to kill and devour them.

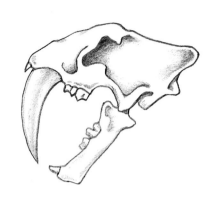

The skull of the *Machairodontinae*, the sabertoothed cat.

The skull of the *Felinae*, with its conical teeth, inherited by today's cats.

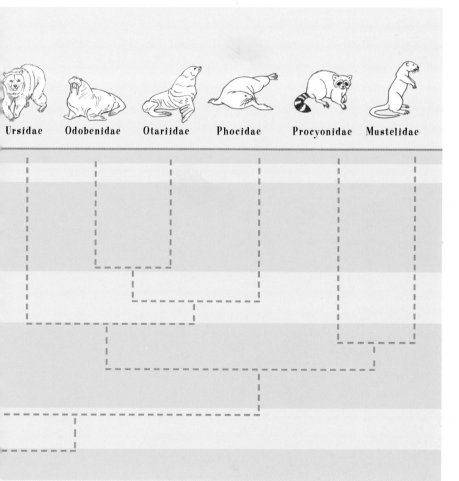

Ursidae   Odobenidae   Otariidae   Phocidae   Procyonidae   Mustelidae

9

The second branch, the **Felinae**, possessed strong conical canines ideal for attacking and seizing prey, as does every cat great and small padding silently over the earth today.

The marvelous and exotic sabertoothed cats, who finally became extinct around one-and-a-half-million years ago, are so familiar to us that it seems strange to imagine that they are not our quick-witted pets' ancestors. However, all the evidence points to the fact that *Felis catus*, or the household tiger as many like to think of her, is descended from the *Felinae*.

Work by Stephen O'Brien, a biologist who uses a system known as the molecular clock (which measures blood serum albumin immunological distances between different species to time the emergence of various cat lineages), demonstrated that the first cats to branch off from the *Felinae*, around twelve million years ago, were small South American cats. These were followed two to four million years later by ocelots and, finally, the lineage to which *Felis catus* belongs.

Of these, Pallas's cat (*Felis manu*) is the oldest member—she lacks the endogenous retroviral gene RD 114, which all the later members of the *Felis catus* family possess. Six to eight million years ago she was followed by the Blackfooted cat (*Felis nigripes*), who is confined to southern Africa; the Jungle cat (*Felis chaus*), who ranges from Egypt through the near east across to India and down to south-western China; the Sand cat (*Felis margarita*), desert denizen of northern Africa and the Near East; and *Felis bieti*, the Chinese desert cat, who actually lives in steppes and mountain terrain. Most recently, the *Felis silvestris* family developed, which consists of three geographic races: the European wild cat, generally known as *Felis silvestris*, and her African and Asian cousins, who, when they were named as species were known as *Felis libyca* and *Felis ornata*—names which are still often used to denote which wild cat is under discussion. It was to be around another four million years before the pantherine lineage, which includes cats such as the cheetah, diverged, and it was a mere two million years ago that what we term as the big cats—lion, tiger, leopard, and jaguar—first paced the earth.

Lions, tigers, leopards, and jaguars first paced the earth two million years ago.

Adapting to Survive

The appearance of *Homo sapiens* varies immensely depending on geography and natural climatic habitat. The eyes of the Inuit people are protected by thick fatty folds that insulate the eye against freezing, and protect them from the constant glare of bright light reflected from snow. The skins of humans who originated in hot, humid, equatorial climates are melanin-rich, meaning they contain dark pigment that helps protect their owners from the damaging effects of ultraviolet light. Cats, big and small, are no different.

Lighter that her Siberian counterpart, the Bengal tiger also has shorter fur to cope with the warm Indian weather.

The cats that are currently termed as different species—although it seems all are capable of interbreeding—are in essence geographical races who have evolved specific characteristics, physiological, hormonal, physical, and behavioral, to exploit to the full their terrain, the circumstances of their environment, and their chances of survival. The quality of light, the density of jungle, the heat of the desert sand, humidity, the presence or absence of predators, including man, all play their part. The tiger, supreme predator of the east, personifies this adaptation. Tigers that inhabit the icy northern latitudes ranging through Siberia are large in body to conserve heat; they weigh in at an enormous 660 lbs (300 kg), have long fur which is white on their chest and belly, while their bodies are marked by brown widely-spaced stripes on a pale amber base. As the tiger's range moves south to the tropics, her body size decreases, the more readily to dissipate heat. The Bengal tiger averages 480 lbs (218 kg), with short fine fur, her coloring rich and dark for camouflage in the dappled light of forests; the Sumatran tiger, living in the equator's hot humidity, has the most melanin-rich and vivid pelage of all, with black stripes contrasting with an intense orange background. Her weight? A mere 260 lbs (118 kg).

Cats of the Wild

Pallas's Cat

Of all the Domestic Cat Lineage Members (DCLMs), Pallas's cat is the most specialized, preeminently adapted to the punishing terrain that is her ecological niche: The rocky, altitudinous bleak and barren steppes and uplands of central Asia. Strange and fantastic looking, Pallas's cat has an unusually broad head, low foreheads and flattened face not unlike that of a Persian cat or Pekinese, while her eyes, unlike those of other DCLMs, close to a circle, not a slit. It is, however, her low-slung ears that allow her to peer over low scrubby vegetation and small rocks without revealing more than a slither of her head or losing the sharp hearing so vital in this landscape. Her short stocky little legs, according to one observer, allow her "not to overtly leap from ledge to ledge but instead appear to 'flow' from perch to perch." She is a mistress of camouflage: Her pale shaggy fur, a silvery iron-gray, make her virtually invisible even when in plain view. The fur is much longer on her throat, chest, belly, and thighs,

The oldest member of the domestic cat lineage:
Pallas's cat.

which protect her vulnerable flesh from the icy unforgiving ground. Pallas's cat, at least in contemporary times, is elusive, her habits unknowable; but according to the great Indian naturalist Prater, in captivity she showed "no fear of spectators nor a desire to avoid them and was very silent, never uttering the familiar snarling growl or hiss." These human-tolerant ways would theoretically have inclined her to approach human settlements in search of rodents, where she might have mated with domestic cats or, because of her unusual but charming looks, been taken as a pet. It is certainly tempting to imagine that she is in some way related to the great tribe of Persian cats, the only other felines to possess flattened faces and noses. The Pallas's cat's protected status is uncertain. Although meant to be protected by signatories to the Convention on International Trade in Endangered Species (CITES), which include China, Afghanistan, Russia, and Mongolia, in reality, lack of funds, and the lack of political will, leave her vulnerable to trappers. Thousands of skins are traded every year. She is a rare cat, "a wise and handsome old character," her numbers unknowable, but not great, and illegal activity is seriously damaging her survival.

Evolved to cope with the extremes of heat and cold in the Middle East, the Sand cat boasts furry toes that protect her paws from hot sand.

The Sand Cat

Asia and the Middle East's Sand cat has had to evolve to cope with burning desert sands where surface temperature can exceed a coruscating 176°F (80°C). Long dense hair grows between her toes, forming a thick pad that both insulates her feet and allows her to walk easily over the sand's fine ever-shifting surface. Her gorgeous thick coat, for which she is so frequently killed, protects her from extremes of temperature ranging from 13°F (-25°C) to 104°F (40°C) and her physiology is such that she is able to extract what moisture she needs from her prey. Like Pallas's cat, the Sand cat has claims to being the Persian's ancestor. She is tractable and docile in the extreme. Combined with her beauty, this makes her a prime candidate for being tamed, and perhaps bred. Her geographic spread coincides with that of the long-haired Persians and both have a mass of dense hair covering their feet. Shorthaired Persians, however, do not and the conclusion

An evolutionary success, the Sand cat extracts water from her prey, only rarely actually drinking.

of zoologist Hemmer was that if she were an ancestor, all Persians, short- and longhaired, would have pads. Of course, it might be that selective breeding could produce shorthaired Sand cats without pads, but we humans are unlikely to find out until both the Sand cat and the Pallas's cat have been subject to genetic analysis. The Sand cat's status in the wild is essentially unknown, but after she was discovered in Pakistan in 1966 she has been hunted by fur dealers and her numbers have plummeted. In Israel her desert domain has been cultivated, making survival impossible, but she still flourishes in Muslim areas of the Sahara. Tradition has it that she, eagle owls, and hoopoe birds were the companions of the Prophet Mohammed and his daughter Fatima, so these animals are free from persecution. (Mohammed was particularly fond of cats. He is said to have cut off the sleeve of his gown rather than disturb a sleeping cat, and to have given the Egyptian cat the streaks on her fur by

The Sand cat (behind) and the Persian cat.

stroking, which bestows protection on even the lowliest of street cats.) In ancient Egypt it was quite commonplace to donate money for strays: The best-known cat charity of the time was that of Sultan Baibars (AD1260–1277) who left a garden to destitute Cairo cats, whose descendants may still be seen lazing in city mosques.

The Chinese Desert Cat

On one of the very few sightings of the Chinese desert cat back in 1923, she proved herself to be more than a match for the fox hound companion of explorer Dr. Hugo Weigold. The hound pursued the cat through low thickets on the mountains east of Sungpan, China. The dog returned alone with two bites on his jaw. The cat punished the dog further the next day, but fled when Weigold attempted to imprison her.

The Blackfooted Cat

The Blackfooted cat lives in a relatively small area of southern Africa. Such a ferocious predator that she ambushes every 30 minutes when out hunting, she succeeds 60 per cent of the time—by comparison the tiger's strike rate is just one in ten. The numbers of this rare species are unknown, but because the cats are tiny—even the males weigh only 4½lbs (2.5 kg)—and do not prey on farm animals, they are rarely persecuted.

The Jungle Cat

Ranging through Egypt, across to India and southwestern China, the Jungle cat is a superb hunter who can run in bursts of up to 20 miles an hour (32km/h). Although she can adapt to desert and steppes, she is a keen swimmer and is often to be found in thick brush reed and teeming swamp where she picks off ducks and water fowl. In India, she favors woodland, plain, and scrub land, but whatever her habitat she opportunistically takes over the abandoned burrows of other creatures to use as her den. Whether this fabulous cat ever became domesticated remains a mystery but she was no doubt a helpmate.

Cairo's population gave charity to protect cats in ancient times.

15

How Cats Became Domesticated

The generally accepted theory is that members of the African race of the wild cat *Felis silvestris*, probably together with the Jungle cat *Felis chaus*, were first domesticated in ancient Egypt, from whence their burgeoning population spread across Africa. The cat's world domination was assisted by mankind who transported this amiable companion to the four corners of the globe. One of her final destinations was Australasia, where she decimated the native wildlife which had no defense against this novel predator.

Cats were invaluable to China's silk industry; five thousand years ago, they were used to dispatch the rats that themselves fed on the silk-spinning caterpillars living on the leaves of the mulberry tree.

It does seem strange that wild felines, who were so widespread, were thought to be domesticated only in Egypt; and certainly some evidence points to China as another likely area. The Chinese cat domestication program was highly advanced more than five thousand years ago. A lap-dog cult flourished in the Imperial court and the emperors kept "dog books" in which the portraits of Pekingese were painted by the leading artists of the day, which established the desirable breed traits of the time. As in Egypt, China had a thriving agricultural system, which would naturally have attracted plump greedy rodents and their attendant feline predators who, according to the Chinese equivalent of the *Encyclopedia Britannica*, soon became objects of reverence. Their silk industry was also established at least five thousand years ago and cats were considered invaluable in dispatching the rats that preyed on the industry's caterpillars. (The caterpillars produced the invaluable silken threads as a byproduct of munching on the leaves of the mulberry tree.) In fact cats became known as "protectors of silkworms" and even their image pinned on a wall as a talisman was believed to keep the valuable silkworms from harm. It thus seems extraordinary that the Chinese would not have tried to domesticate a creature so important to their food supply and economy. The encyclopedia states categorically that the Chinese, not the Egyptians, were the first to domesticate the cat, and cites records from as early as the Western Zhou period (1766–770BC). The ancient Chinese *Yi King* or *Book of Songs* which dates from around 800–600BC (but is probably based on even earlier work), mentions wild cats being hunted for their fur. Confucius (551–479BC) apparently kept a pet feline, as this anecdote attests:

"Tseng Shen and Min Tzu, canonised disciples of the Sage, were listening outside to the music of the Master, who, as was his wont, was soothing himself by the lute that he loved so well, when suddenly the strain changed. On entering and inquiring what the change meant they were told by Confucius that he had seen a cat making for a rat, and that he had struck up another tune to stimulate the cat in its attack upon the rodent."

Certainly, by AD725, the domestic cat was so established and numerous in China that she was taken to be used in medical research, as she continues to be worldwide. One such report of a medical experiment tells how "the prolonged consumption of rice weakens the body" and that "feeding polished glutinous rice to young cats and dogs bends their legs so they are unable to walk." This, although the researchers did not realize it at the time, was proof of the symptoms of the malnutrition disease beri-beri.

The African wild cat, ancestor of today's household pet.

The Cat Reaches Japan

Whether domesticated independently or imported from Africa, what seems certain is that the Chinese introduced the cat to Japan during the reign of Emperor Ichigo (AD986–1011). By AD999 the imperial cats were established and breeding and "the left and right ministers had the task of bringing up the kittens and prepared boxes [with delicacies] and rice and clothes for them [as for newborn babies]. Uma no Myobu was appointed wet nurse for the kittens. The people laughed at the matter and were rather astonished." This however did not stop the emperor promoting his favorite felines to the the fifth rank—that of lady in waiting.

The beckoning cat is one of Japan's favored greeting motifs even in contemporary times.

Adored by the nobility as extremely desirable and expensive status symbols, cats were frequently called by the pet name *tama*, or "jewel," and taken for walks on a leash.

In 1602 a plague of rats and mice threatened the Japanese silk industry, and the cat was urgently required in her professional capacity so that disaster might be averted. The Kyoto authorities placed placards ordering, "Firstly the cords on the cats in Kyoto shall be untied and the cats set loose. Secondly, it is no longer allowed to buy or sell cats. Whosoever transgresses this ordinance shall be punished with a heavy fine."

Why the common people's cats were tied in the first place is difficult to determine. It may have been because they were still valuable and relatively rare or even because they were considered dangerous in some way. Certainly she was commonly associated with hauntings and demonic activities by everyday people. This tale from Echigo, written in 1708, is typical:

"Every night the samurai's house was visited by luminous balls, impossible to catch, which flew across the floor and illuminated a tree in a neighbor's garden. The maidservants were also subject to haunting—spinning wheels turned of their own volition and one maid was unable to sleep as her pillow moved in all directions. In vain help was sought from sorceresses, Shinto & Buddhist priests and *yamabushi*. Then, one evening, the samurai was strolling in his garden and spied a very old cat wearing his maidservant's towel on its head, walking along the roof and 'anxiously looking about with its paw above its eyes.' The 'brute' was killed with an arrow and the haunting ceased."

Western powers were never able to colonize the fiercely isolationist Japan which meant neither could the otherwise ubiquitous blotched tabby. The cats of Japan developed in their own unique way from a limited gene pool whose genetic mutations gave rise to the beautiful, natural pom-pom-tailed feline known as the bobtail and the auspicious tricolored red, black, and white cat known as the *mi-ke*. The Japanese artist Utagawa Kuniyoshi (1797–1861) immortalized them all in his charming triptych of the various cats from the 53 post-stations of the Tokaido road. In Japan now cats remain as varied, attractive, and healthy as they did then, causing Roger Tabor to write of a visit to Tokyo in the 1990's, "watching them play was to see Kuniyoshi's sketches come to life."

The bobtail is considered auspicious and is often rendered in sentimental ceramic style as a manek-neko, or beckoning cat, with one paw raised in invitation—making them a must for store and restaurant owners. The bobtail seems to suffer from problems linked to the recessive gene which creates its unique tail, but her genetic inheritance also makes her a robust, very sociable, and extremely talkative cat. Her voice covers a whole scale of tones and according to the Cat Fanciers' Association she has even been known to sing. She also "enjoys a good game of fetch," which should make her exceptionally popular with those who favor a dog-like cat.

The auspicious bobtailed cat.

Clues to the Early Cats

Whatever the secrets of domesticity prove to be, the earliest-known cat skeleton dates back to before 4000BC. It was found buried at the feet of a craftsman in the ancient cemetery of Mostagedda, of Middle Egypt. A gazelle, also buried with the man, would have been to provide the man with food as he journeyed in the afterlife; but the cat, having no food value for Egyptians, seems likely to have accompanied him as his pet—was she a domesticated feline or a tamed wild cat?

We cannot know for sure because such recent domestication means that the skeletons of wild cats and domestic cats are still so similar that it is impossible to tell them apart.

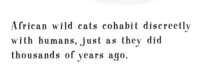

African wild cats cohabit discreetly with humans, just as they did thousands of years ago.

Longer periods of domestication usually lead to quite distinct morphological changes, a process which was demonstrated in Siberia where wild silver foxes were bred selectively for just one characteristic—tameness. The overwhelming majority of the first-generation wild foxes either fled from their handlers, bit them, or, when stroked or handled, resolutely ignored them. After just six generations of breeding for the smallest signs of affection, fox cubs were being born who were actively interested in communicating with people. With this behavioral shift came neurochemical and hormonal changes, including a steady drop over the generations of hormone production, including epinephrine from the adrenal gland. This crucial change meant that the foxes were not stressed by human contact. This also happened when wolves, over tens of thousands of years, gradually morphed into the domestic dog and these changes were further enhanced by man's selective breeding, giving us the huge variety in canine form we see today. The cat, at most, has been domesticated for only six thousand years and has never been bred for specific behavioral traits in the same way the dog (or the cow) have been.

Some wild cat species may lack the fear of man that is present in most other wild creatures. For the Siberian foxes to become real pets, which they finally did, behaving in every way like the domestic dog, the output of their adrenal gland had to reduce by 75 per cent. This may not be the case with certain wild cats, and if their neurochemical and neurohormonal systems did not need to change, or needed to change very little for them to coexist happily with us, then common sense dictates that their bodies would also remain more or less unchanged. Today African wild cats, just as they no doubt did thousands of years ago, happily cohabit with human companions and frequently live close to human settlements, feasting on fat greedy rodents or patroling corn plantations where rats are also plentiful. Reay Smithers, a zoologist living in Botswana during the 1960's, kept three wild cats as pets who had the run of his house and large garden enclosures. Just as our pets do, Komani and Gori shared the sofa, walked over any paper Smithers happened to be writing, and insinuated themselves between his face and what he was reading, demanding attention. Initially Smithers made the error of releasing his rather territorial females Gori and Komani together just as he was going on holiday. Komani was roundly dispatched from the compound by her competitor and spent four months in the bush. Smithers and his wife looked for her constantly and were finally rewarded. Clearly her return was dictated by her affection for Smithers rather than need.

Zoologist Reay Smithers noted that his cats were in better condition when they lived wild but returned to him and his wife out of sheer affection.

"I knew this night it was Komani because instead of moving off as the others [cats] had done, she very hesitantly moved towards me—near but not near enough to catch. I called my wife to whom she is particularly attached, and we sat down while she softly called the cat's name. It must have taken a quarter of an hour before Komani responded and came towards her. The reunion was most moving, Komani going into transports of purring and rubbing herself against my wife's legs. It was to some extent humbling to find her in better condition than she had ever been under our immediate care."

Cats in Ancient Egypt

A courageous feline, the African wild cat does not hesitate to take on the cobra and, when unable to best it, fight until mutual death is the outcome.

In some ways the ancient Egyptians did not distinguish between gods, people, or animals. All were classified as "living beings" and, perhaps unsurprisingly, they further failed to distinguish between wild and domestic cats in their early hieroglyphics, simply calling them "miu," "mii," or later "imi," all of which mean "[s]he who mews." All this means that signs of feline domestication in Egypt have to be gleaned from context and even then, early on, there are mysteries to be solved.

The ancient Egyptians had two important natural enemies: Poisonous snakes such as the cobra or the horned viper whose bites were deadly, and voracious grain-gobbling rodents. The African wild cat makes short work of these larger snakes, while the Jungle cat is adept at dispatching smaller reptiles such as racers, whips, and scorpions. A courageous feline does not hesitate to take

The Sacred City of Miuu

A fragment of a limestone temple wall near the pyramid of King Amenemhat I (1980–1951BC) 30 miles (48 km) south of Cairo contains a scene depicting an anonymous deity known only as "Lord of the city Miuu." The three hieroglyphic cats that make up the word Miuu are perhaps the oldest "domestic" cat images so far uncovered, and as stylistic evidence points to some of the limestone blocks being taken from earlier dilapidated structures, it is just

possible the image dates back to 2200BC. They may indicate that the area was called "cats" or that this was "cat town," but as hieroglyphs could be used only to indicate pronunciation, the word may just be "Miuu" and have no meaning that we can discern. Records reveal there was a town with a similar name around Thebes that "would have been connected with a genuine cattery in a local temple precinct," so the balance of probability is that it really did mean "cats."

on the cobra and, when unable to beat it, will fight until mutual death is the outcome. Clearly the cat—wild, tame, or domestic— would have been enthusiastically welcomed by these early agriculturalists and the luxury-loving, sensuous feline would, in her turn, have found it advantageous to cohabit with man, sharing in all the benefits of his civilization.

An early indication of the reverence with which the cat would be held in later centuries is shown by a spell written on an Egyptian wooden coffin dating back to 2100BC which mentions a great tom cat and asks "Who is this great Tom cat? He is the god Ra himself. He was called cat [Miu] when Sia [the personification of knowledge] spoke of him because he [the cat] was mewing during what he [Sia] was doing and that was how his name of 'cat' came into being." About this time magic knives engraved not only with the cat's image but that of other important creatures, such as the crocodile, began to be used as talismans to invoke protection against the vicissitudes of life, both real, and imagined, and seem to have been extremely popular with ladies and children.

The eternal nature of the cat, after an Egyptian tomb painting.

The Ancient Egyptian Cult of the Cat

Animals have always been a vital element of Egyptian cosmology and two anthropomorphic deities, the lion-headed Sekhemet and her sister the cat-headed Bastet, represented the fascinating and powerful feline.

As lions were commonplace in Egypt four thousand years ago, Sekhemet's places of worship were often sited on the boundary between wild desert, the big cats' natural habitat, and the civilized, populous Nile Delta. Associated with powerful energies such as wrath and vengeance, Sekhemet employed lions against enemies of the sun god Ra.

Bastet, originally also lion-headed, had become a benign cat-headed deity associated with fertility, fecundity, and maternity by the time of the New Kingdom (1570–1070BC). One of Ra's all-seeing eyes, perhaps in acknowledgment of the flesh-and-blood cats' keen sight, Bastet was often represented with a stylistic eye which in amulet form was a powerful charm against evil.

Herodotus, the Greek historian, visited Bastet's temple at Bubastis in 450BC describing it as the most beautiful in all Egypt. Constructed of fine red granite and surrounded by elegant whispering trees and wide canals fed by the great River Nile it must have been a marvel of its time. He describes how many devotees clearly took Bastet's association with pleasure, love, and joy to heart as they made their yearly pilgrimage to her sacred shrine: "Men and women come sailing all together, vast numbers in each boat, many of the women with castanets, which they strike, while some of the men pipe during the whole time of the voyage; the remainder of the voyagers, male and female, sing the while, and make a clapping with their hands. When they arrive opposite any of the towns upon the banks of the stream, they approach the shore, and, while some of the women continue to play and sing, others call aloud to the females of the place and load them with abuse, while a certain number dance, and some standing up uncover themselves. After proceeding in this way all along the river-course, they reach Bubastis, where they celebrate the feast with abundant sacrifices. More grape-wine is consumed at this festival than in all the rest of the year besides. The number of those who attend, counting only the men and women and omiting the children, amounts, according to the native reports, to seven hundred thousand."

Thousands of cats, tended by Bastet's priests, lived luxuriously in the grounds of her temple. Those who could afford to send the bodies of their own precious pets to Bubastis to be buried, imbued with costly oils and aromatic scents, in her sacred repositories or one of the other underground cat cemeteries constructed along the banks of the Nile. In 1889 300,000 mummified cats were discovered at the Beni Hassan site alone.

Embalming was particularly important to the Egyptians as they believed living creatures had a soul, or *ka*, which lived on after the material body died, and would one day be resurrected. This *ka* needed to be the double of something—the actual body was best— hence mummification. Should the mummy be destroyed the *ka* could also survive in little statues or portraits of the dead.

"A rich man's cat was elaborately mummified, wound round and round with stuff, and cunningly plaited with linen ribbons dyed two different colors. His head was encased in a rough kind of papier mâché and that was covered with linen and painted, even gilt sometimes, the ears always carefully pricked up. The mummy might be enclosed in a bronze box with a bronze Ka statue of the cat seated on the top." Some also had eyes of obsidian or rock crystal inlaid into the papier mâché.

In life the ancient Egyptians treated their cats with every courtesy and, like many contemporary cat devotees, adorned them with expensive jewels and silver chains. In death households went into elaborate mourning and the cat's companions shaved off their eyebrows as a mark of respect. To kill a cat was considered a particularly heinous crime punishable by death.

The Egyptians so revered their cats that they dressed them in jewels and anointed them with perfume.

Cats in Tomb Paintings

A domestic feline image is first shown in the paintings that surround Egyptian official Baket III in his rock-cut chapel at Beni Hassan, dated 1950BC. Beside colorful renditions of four baboons, who were also extremely popular pets at the time, and a man holding a stick "reminiscent of an instrument often seen in the hands of house attendants who look after pets," a cat is shown dispatching a rat.

Diodorus Siculus, an ancient Greek historian who traveled in Egypt, confirms that pets were even then being tempted with chunks of raw fish; but that another prized dish was bread and milk.

But a definitive indication of the cat's ever-escalating ascendancy is the cat chapel at Thebes's Abydos cemetery (1980–1801BC) where 17 cat skeletons and a little row of offering pots, which had probably contained milk to sustain the cats through the afterlife, were discovered. Since around 2500BC Theban tomb paintings of men showed either a monkey—a symbol of masculine virility—or faithful canine. A thousand years later their wives claimed the cat, whose image no doubt functioned both as a symbol of feminine fertility, as a pet. Egyptian artists felt leaving large unused spaces went against good composition, so many of even the most fetching of pet felines may only have been figments of the painter's imagination. Most of the Theban felines are depicted as busily engaged in eating—everything from fish as large as themselves to huge animal legs. Diodorus Siculus, a Greek historian who traveled to Egypt in 60–57BC confirms that pets were still being tempted with chunks of raw fish but that another favorite dish was bread and milk. Where the once vivid paints of the Egyptian artists are still visible, the colors—between sandy yellow and light-reddish brown—and the patterns (similar to those of striped tabbies) echo those of the friendly African wild cat.

Many beautiful and evocative paintings of felines in marshland hunting scenes also exist. Throughout the east the cheetah has been tamed and used to hunt game—Mughal rulers kept stables of upward of a thousand of these svelte predators, so there seems no reason that the Egyptians should not have employed indigenous cats for the same purpose. The Jungle cat, like the cheetah, is a superb hunter and, as a hunting scene from the tomb of Nebamun at Thebes *c.*1400BC shows a cat with the distinctive three rings on her tail, this may be evidence of the Jungle cat being used to flush out and retrieve prey. Further indications that the cat was at least tamed, if not domesticated, come from mummified remains, although most feline mummies are those of African wild cats.

The Earliest Comedy Cats

Although much is made of the cat's deification—and the cat cult which surrounded the goddess Bastet whose temple at Bubastis (at Tell Basta, 25 miles (40 km) northeast of Cairo) contained a cattery of thousands who were attended by the goddess's priests— it is the humorous representations of her which really bring home how much an accepted part of everyday life the cat had become. These images were created by the skilled artisans working on the construction and decoration of the pyramids of the pharaohs of the New Kingdom at Deir el-Medina. These satirical papyruses and sketches on white limestone, which date back to 1500BC, were created entirely for the artists' own amusement. Wishing to bring the mighty cat down a peg or two, in these works it is the rats and mice who rule while cats attend to their every need. Slavelike felines fan finely-dressed lady rats, while whole retinues attend to elaborate rodent coiffure, nurse baby mice, and provide succulent tidbits.

Slave cats attend to ruler rats in an ancient Egyptian "comic strip."

CHAPTER TWO:

CAT BREEDS AND BEHAVIOR

"Of all animals, the cat alone
attains to the contemplative
life."

—Andrew Lang (1844–1912)

The color of luxuriant fur, the blaze of an
eye-green, gold, or blue-are the only things
that distinguish all but the most venerable
breeds of cat, notably the Persians and the
Siamese, from one another. Cats display a
range of temperament, disposition, and
character that rivals humans; with their
almost magical variety, we can glean only
a few clues to the delightful, shocking, or
appealing traits a cat may exhibit as a
result solely of her breed.

Ancient Breeds

The most ancient of felines, at least in her homeland, must surely be the lithe and exquisitely marked Egyptian Mau who, whether a street cat in Cairo or a pampered pedigree, often resembles so entirely the Theban tomb cats she might be a flesh-and-blood relic. So striking as to be breathtaking, it is no wonder she was avidly worshiped and adored in ancient Egypt. Her genetic diversity is such that felines of every character, from the trusting and sweet to the proud and aloof, patrol the streets and alleys of her native land, but how the inbred Maus of the US and the UK will develop is yet to be seen.

The spotted Egyptian Mau, once worshiped in Egypt.

That the domestic cats of Egypt are the sole ancestor of *Felis catus* throughout the world is still open to question, but certainly the Roman Empire's close trading links with Egypt make it more than probable that the gorgeous svelte Mau was transported to Italy. "How silently and with how light a tread do cats creep up to birds, how stealthily they watch their chance to pounce on tiny mice," wrote Pliny, confirming his intimate relationship with at least one cat. And yet, other than this description, there are only a handful of feline images in all the rich remains of the Roman Empire; and the ruins of Pompeii, have not, so far, yielded one cat skeleton or larva-formed mold of their decayed corpses. There are, of course, many records of animals fleeing cities or behaving curiously prior to earthquakes and other natural disasters, and signs of Vesuvius's imminent eruption must have been clear to feline and man alike—but if pet cats were commonplace, it seems curious that none were caught in that fatal lava flow.

The European Wildcat

Many people believe that the Roman Legions were
responsible for bringing the domestic cat to Great
Britain, but this does not preclude the separate
domestication of the European wildcat—not only
in Great Britain but across the entirety of her
range throughout Europe to Moldavia and parts
of Russia.

Her skeleton is present in every prehistoric site
where bones are found in quantity and often in
association with those of man. As the hunter-
gatherer way of life was gradually replaced by
farming throughout Europe, it seems reasonable to
assume that the European wildcat would have taken
advantage of the rodent bounty that grain brings and be
welcomed by man for the same reasons as her African cousin was.
The major argument employed against this is the European
wildcat's undisputedly fierce, combative, human-hostile nature.

Many people believe that the Roman
legions and their armies were
responsible for bringing the
domesticat cat to Britain,

Bingley, a naturalist writing in 1813, described her as the "British
Tiger" and wrote: "These animals are sometimes caught in traps,
and sometimes killed with the gun. It is however, dangerous to
merely wound them, for in this case they have frequently been
known to attack the assailant; and their strength is so great as to
render them no despicable enemy."

Her essence is further embodied perfectly in this description of a
captive, shown at the first cat show at London's Crystal Palace,
written by its founder, Harrison Weir, in 1871. The victim had
been caught in a trap on the Duke of Sutherland's estate and its
foreleg injured, but "not so as to prevent it moving with great
alacrity, endeavouring frequently to use the claws of both forefeet
with a desperate determination and great vigour. It was a powerful
animal possessing great strength, taking size into consideration, and
of extraordinary fierceness." It refused to leave its thick and heavily
barred traveling box and "maintained its position, sullenly retiring
to one corner of its box, where it scowled, growled and fought in a

most fearful and courageous manner during the time of its exhibition, never once relaxing its savage watchfulness or attempts to injure even those who fed it. I never saw anything more unremittingly ferocious nor apparently untameable." It was, Weir continued: "Bolder [than the domestic cat], having a rugged sturdiness, being stronger and broader built, with forearms thick, massive, and endowed with great power, with long curved claws. The feet were stout, sinewy and strong."

Over the past millennium there has been very good reason for the savage demeanor of the European wildcat, the tree-climbing terror of dense forest and vegetation-her intense persecution by human predators.

There is, however, at least over the last thousand years or more, good reason for the savage demeanor of this tree-climbing denizen of forest and dense vegetation— her intense persecution. Top predators, such as the tiger, become increasingly belligerent and aggressive toward man in direct relationship to the degree of their oppression and abuse, and the European wildcat has been designated as vermin throughout much of her habitat for centuries.

In the Middle Ages the wildcat was aggressively hunted for her fur by peasants; it was designated as sufficiently lacking in kudos to be suitable apparel for nuns. In 1127 Archbishop Corboyle decreed that "no abbess or nun use more costly apparel than such as is made of lambs' or cats' skins." And with the emergence of the great estates she was shot and trapped mercilessly by gamekeepers, who considered her to be a major predator of pheasant, grouse, and rabbit. By 1881 she was on the verge of extinction in the British Isles, her last pawhold, then, as now, being in the deer forests of northern Scotland. The cats who survived from necessity would have been crafty, intelligent, gladiatorial, rugged, tough, and strong, but they may well have been far more tractable two thousand years previously.

Although the Egyptian's pet breed has clearly contributed to domestic stock throughout Europe, and the European wildcat has been interbreeding with both domestic and feral cats for at least two thousand years, it is truly hard to imagine, when a contemporary mackerel-colored tabby is placed next to a European wildcat, that she is not her direct descendant. Anecdotally, Weir wrote that in his experience when cats returned to their feral state, in just a few generations they reverted to their original wild form. He thought that a natural and nutritious diet and the physical demands put upon them selectively increased their size and muscle bulk, while a propensity for pale-furred cats to be shot favored the survival of the darker-striped markings typical of the European wildcat. Whatever the truth the domestic cats of Europe have extraordinary genetic diversity. Researchers Randi and Ragni found that Italian domestic cats possessed more diversity than either African or European wildcats and that this huge range was mirrored in cats that other scientists had studied elsewhere— further evidence that, be they blotched tabbies or marmalade toms, no inferences can be made about their personalities from their attractive good looks.

This turns, however, to a somewhat different story when we come to look at the ancient breeds of the East.

The Siamese Cat

Many people, breeders included, are not aware of their gorgeous felines' venerable pedigrees. Before feline documentary maker Roger Tabor showed his series on cats, which featured in part the history of the Korat (Si-Sawaat), one of the few cats in the West that still resembles its ancestors to this day, he was simply amazed to hear breeders exclaim: "What a pity there is no such place as Korat!" In fact it is a huge plateau occupying the greater part of eastern Thailand, where the Korats, with their bewitching heartshaped faces, gleaming green-yellow eyes and thick, dark iron-gray fur can be seen to this day. Although Siamese cats are feted in the west, it is the Si-Sawaat (who for centuries been revered as a bringer of rain to the dry drought-worn province that bears her name) with which the Thais more closely identify themselves.

Cats have been bred in Thailand for hundreds of years, something Westerners are unaware of because the kingdom was closed to travelers until the middle of the nineteenth century, and because most of the ancient traditional Thai books, or *samut khoi*—

Unknown to the West, cats have been bred in Thailand for hundreds of years in the once-hidden kingdom.

The "cat of the diamond eyes" has a reality sadly rooted in cat flu, which produces glittering pupils.

manufactured from khoi tree pulp into one long sheet and folded concertina fashion—were destroyed when the Burmese sacked the ancient capital of Ayuttaya.

The Siamese Cat Treatises

Despite early Thai books being extremely vulnerable to humidity, rats, and insect predation, a handful of documents known as *Tamra Maew*—the classic Siamese Cat Treatises—still survive in museums. Dating from the eighteenth and nineteenth centuries, but based on an earlier tradition going back to at least the seventeenth century, these Treatises have been translated by Martin R. Clutterbuck, a Thai scholar and catophile. He has revealed them to be the world's oldest breed standards for no less than seventeen distinct feline races. Tabbies receive only one short mention under "bad cats," indicating that the first cats to reach Siam in prehistoric times must, unusually, not have been the regular African wildcat, or her domestic progeny, but a group of felines who had undergone a spontaneous coat mutation. Constant breeding between these first cats would have produced races with their own genetic identity, leading to looks and perhaps character at variance to the host of cats directly domesticated from the beautifully striped wildcats.

Some of the seventeen ancient breeds are now extraordinarily popular in the West. The *Maew Kaew*, for example, is white with black ears, mask, paws, and tail, better known to us as the Siamese point; then there is the *Dork Lao* or Korat; while others such as the *Wilat* or "Beauty" (see box, opposite) and *Ratanakampol* or "Jeweled Cloth" seem to have completely disappeared.

Although at first glance it seems outlandish to imagine there was ever a breed of cats with a thick dark stripe around the middle of its body (see box, above left) in fact this configuration occurs in various animals, including the domestic pig.

Another truly fascinating cat included in the slightly more modern Astrology Treatise *Prommachaat Chabab Somboon* is "she of the Diamond-Eyes:" Clutterbuck describes her in glowing terms. This

"Gray cat with small partly closed eyes. The cat of the five-fold millionaire. By day it goes out and will not raise its head to catch food. Merely staring its victim will fall down. By night it has mucus in its eyes. They say its eyes are gems of massive value, khot taa maew. Having it brings virtue. They are one in ten thousand, and very rare. This is the best cat."

The Siamese Cat Treatises: Wilat

"Round from throat and underbelly, two ears … White to the tail— cotton flower … All four white paws—two green eyes … The name of Beauty for the bodies' black field."

also seemingly fantastic cat is actually rooted firmly in reality. Cat flu—feline viral rhinotracheitis and feline calcivirus—sometimes leads to a condition where the cat's eyes permanently mist over with mucus but, just occasionally the mucus can clear, leaving the eyes jewel like and able to refract light. In recent times, this disease afflicted Sirimongkhol, a pure white cat belonging to the abbot of Wat Suanmanisap and indeed it proved to be "the best cat" as when news of this fabulous cat hit the press, the abbot was inundated with offers of help to build a much-needed chapel. Less believably, but in keeping with the Astrology Treatise's description, the abbot reported that "when the cat looked up at a house lizard, it fell down, until there were no lizards left in his quarters."

The Burmese Cat

Clutterbuck speculates that having sacked Ayutthayan, the returning Burmese armies took a number of *Thong Daengs* or Copper cats (one of the Treatises' original 17 breeds) with them, which became the founding stock for the much-feted lustrous, rich brown Burmese temple cat. Full Burmese cats certainly still live wild in the temples of Bangkok and, under their Thai name *Thong Daeng*, are also a popular pedigree breed. In the West, the term Burmese is, however, something of a misnomer as the race stems from the mating of Wong Mau, a walnut-brown female native of Rangoon, who, on her arrival in the US in 1930 was promptly mated with a Siamese.

The Siamese-whose name in Thai means "jewel" or possibly "moon diamond"-still thrives today.

Are Pedigree Cats Less Aloof?

Little research on communication and feline breed differences has been carried out but Denis Taylor, founder of the Institute for Applied Ethology in Switzerland, has studied differences in interactions between humans and moggy, Persian, and Siamese cats. He confirms that Siamese spend significantly more time interacting with their humans than even Persian cats. Both pedigree Siamese and Persians are concluded to be "more predictable" and less "independent" than the moggies; in effect, much more *doglike*.

All the Oriental races, and the Siamese—whose name in Thai means "jewel" or possibly "moon diamond"—in particular, have a reputation for being exceptionally people-orientated, ultra-attention-seeking, extra-talkative, and super-playful. They are decidedly doglike cats, causing one early breeder to note "they can be easily taught to turn back somersaults, and to retrieve, and in the country take long walks like a terrier." One aristocratic Siamese of my acquaintance always made it her business to accompany me on trips to the local store.

But are oriental cats and Siamese in particular, really so different from their western relatives—or is this wishful thinking by enamored owners? In their native Thailand, these, to us exotic, creatures teem through every nook and cranny of the carved stone Buddhist *wats* or temples as they have for centuries. They are welcomed not only because they preserved the sacred Buddhist manuscripts from the predations of rats and mice but because, meditative in aspect, the cat was a congenial companion for those who themselves meditated. Ancient lore also held that it was to the warm body of a cat the soul of truly enlightened individuals flew, only reaching paradise when this earthly vehicle finally died.

Fed, lovingly tended, and living in a harmonious environment, these Orientals have lived in feline sociability with one another, the monks, and thousands upon thousands of temple visitors and supplicants for hundreds of years. Free to live where they chose, those who remained in the temples had perhaps more tractable, and sociable natures than those who left and may, like canines, have had a predilection to fathom the mental workings of their human companions the better to survive. If you like a kind of feline self-domestication took place which, over the

Ancient Eastern lore maintains that truly enlightened souls are reincarnated as cats.

generations, manifested in a desire not just to inhabit the same physical space as humans but to try to communicate actively with them. The Siamese were also considered to be very fussy eaters. In their native land, Siamese cats have always been fed on a delicious diet of grilled fish mixed with cold rice. The *Klon Phleng Yaao* Cat Treatise firmly advises on all aspects of the cat's welfare: "The cat's bed must be kept clean, with food and fish of all kinds provided as appropriate, for the cat's merit is great. They are rare, so hurry to look after, do not suddenly be angry."

Thai treatises maintain that "The cat's bed must be kept clean, with food and fish, of all kinds provided as appropriate, for the cat's merit is great."

A Miss Walker, whose father in the 1860's was the first person allowed to bring Siamese cats to the UK, reported: "We feed them on fresh fish with boiled rice until the two are nearly amalgamated; they also take bread and milk warm—also an ancient Egyptian feline-favourite, the milk having been boiled and this diet seems to suit them better than any other. They also like chicken and game."

Things of course have gone downhill for western cats since then. Now many Siamese are forced to eat crunchy dried food and gooey gourmet sachets to which their particular digestive systems have not yet adapted. No wonder they are fussy.

Persians and Angorans

Persian and Angoran cats have an equally distinguished history and also have predictable behavioral patterns. They have frequently been interbred and are usually grouped together. As they originated in Iran (Persia) and Turkey respectively, and these countries share a border, it is not surprising that they have many similarities as there must have been ample opportunity for gene flow between the two.

Friends of the power behind the throne of seventeenth-century France, Cardinal Richelieu's cats were bequeathed pensions in his will.

The enigmatic Mona Lisa del Gioconda, painted as the *Mona Lisa* by da Vinci between 1503–1506, owned a "white cat of a rare species brought from Asia, whose eyes were of different colours, the right as yellow as topaz, the left blue as sapphire," that amused her during the long hours of sitting. But it was not until the Venetian traveler Pietro della Valle (1586–1652) paused at the mysterious many-domed city of Isfahan that the exotic longhairs firmly established a pawhold in western Europe. He was so captivated by cats with hair six inches long—"silky, lustrous and a blue gray color"—that he purchased four pairs and had them transported to Rome. Turkish cats similarly captured the heart of the early seventeenth-century French traveler Claude Fabri de Pieresc (1580–1637), whose biographer Gassendi wrote, "he procured out of the east, Ash coloured, Dun and Speckled cats, beautiful to behold." The Angora's sixteenth-century debut in the courts of France and England was a success; but the elegant Persian soon became the longhair of choice, the Angora fading into obscurity in Europe. In the 1950's and 1960's interest revived and cats were taken frm Turkey to found the stock of Turkish Angoras. But it was

Persian (left) and Modern Angora (right) cats

not just in Europe these denizens of the Middle East reigned supreme. They also conquered China—so much so that the chronicler Ch'ien Lung (1736–95) recorded that "Peking [Beijing] is noted for its large Persian cats and extremely small Fu Ling dogs," and Abbé Grosier wrote:

"the cat in China, as in Europe, is the tender object of predilection and the favorite of the gentler sex. Those of the province of Pechili have obtained preference over their rivals by their pretty ways and their fine coats. The Chinese ladies never allow them to leave their apartments where the most delicate of nourishment and tenderest of care are lavished upon them. These cats are of a pure white, their coat is very long, the hairs fine and silky. They do not catch mice and leave this ignoble chase to the cats of the ignoble race with which it be noted China is abundantly supplied."

These luxurious foreign invaders were utterly indulged by the Ming emperors; one Imperial eunuch complained that the noise they made was so overpowering that it caused the Imperial offspring to sicken and die, and he hoped passionately that they might be confined to quarters. Oriental longhairs seem to have had very distinct characters long before they were intensively and selectively bred in the twentieth century. Soninni de Manoncour, a French naturalist in Egypt during the latter part of the eighteenth century, lived with an Angora which displayed all the phlegmatic temperament that these cats are credited with. On journeys the cat "reposed tranquilly on the knees of any of the company." Affectionate in the extreme, she licked the hand that stroked her and when Soninni was writing or thinking, she interrupted him with "little caresses, extremely affecting." She possessed, in a word, the disposition of the most amiable dog, beneath the brilliant fur of a cat." Since then longhairs have undergone intense selective breeding, which has modified their bodies and, seemingly, their characters. A survey by vet and author Bruce Fogle revealed that among several breeds, Persians now demanded least attention, were the least excitable, and least playful.

The phlegmatic Persian relaxes, while less sedate moggies frolic at her feet.

Breeding Cats for Show

Harrison Weir, who founded the original cat shows, did so with the most humane of intentions, hoping both that a creature who was then so frequently abused might become cherished by the public at large and that "the different breeds, colours, markings, etc., might be more carefully attended to, and the domestic cat, sitting in front of the fire, would then possess a beauty and an attractiveness to its owner unobserved and unknown because uncultivated heretofore."

A relaxed attitude prevailed at these early shows as evinced by Frances Simpson in her seminal work *The Book of the Cat* (1902), who revealed that although British cats had to be registered at the National Cat Club, they could be entered as "pedigree unknown." She did advise fanciers to make sure of their cat's race as: "It is a grievous disappointment if through ignorance or carelessness a good specimen is labeled 'Wrong Class'," also warning that "cats are such terribly timid shrinking animals that when dragged out of their pens with great difficulty—for the doors are most inconveniently small—they often struggle so violently that for

Two tabby-point Siamese cats.

fear of hurting the animal or of it escaping, the judge will swiftly restore it to its resting place without having obtained much satisfaction from his cursory examination."

Twenty years after founding the cat show, a disillusioned Weir wrote: "I found the principal idea of many of its members consisted not so much in promoting the welfare of cats as of winning prizes. I have left off judging cats because I no longer care to come in contact with such 'lovers of cats'."

Things are now far worse than Weir could ever have imagined, with judges awarding points, not for health, vitality, beauty, and overall robustness, but for exaggerated and extreme body conformations leading, in many races, to the deformity that has afflicted so many pedigree dogs.

The beautiful Siamese has been progressively selected for a slighter and slighter frame and a long, tapering, wedge-shaped head to emphasize its oriental glamor and increase the much desired show-ring "daintiness." This process has led, in the early 1990's, to the breed virtually collapsing due to an overall lack of vigor and increased susceptibility to disease.

Persians have been bred progressively for fuller, flatter faces. Now the Peke-faced or ultra-Persian, as she is known in the US and the UK respectively, is as deformed as the flatfaced snuffling, Pekingese dog. However, even the governing council of the cat fancy in the UK draws the line at the top of the nose leather being higher than the lower lid of the eye.

The beautiful Siamese was progressively selected for a slighter and slighter frame and a long tapering wedge-shaped head to emphasize its oriental glamor.

While Persians become more dependent on their humans, the handsome Maine Coon maintains the genetic diversity and strength that makes her a top ratter.

The Dangers of Selective Breeding

Selective breeding has endowed Persian cats with, among other things, an overwhelming propensity for respiratory, pharyngeal, and eye diseases. These incude conditions such as hereditary entropion, in which inward-curling or loose eyelids cause the eyelashes to rub across the eyeballs, resulting in squinting, perpetually watery eyes, sensitivity to light and, in time, ulceration and scarring of the cornea. A further distressing condition, at least for the cat, is the occurence of feline corneal sequestrum in which dark oval or circular lesions of varying size occur around the center of the cornea, accompanied by a brown or black discharge. Sometimes the lesions slough off, but they can perforate the cornea.

In 2003 research by French vets, using ultrasonographic screening, discovered that polycystic kidney disease, or PKD, an inherited autosomal-dominant disease first detected in the 1960's, was present in 41.8 per cent of Persian cats—a prevalence reflected in pedigree Persians worldwide. PKD is progressive and leads to irreversible renal failure.

Even the Persians' once-strong, thick, lustrous fur, incapable of matting, gloriously adapted for the cold, northerly, and often mountainous regions they inhabited, has been destroyed. Breeders have proudly turned them into "powder puffs" with fine textured silky underhair almost as long as their guard hairs, meaning it mats constantly. One of nature's most fastidious of creatures has been turned, by man, into a freak unable to groom herself. Not suprisingly in Bruce Fogle's cat survey (see page 35), vets rated

LE CHAT D'ANGORA.

An early breed manual recorded the Angora cat as the favorite of Marie Antoinette.

Persians as the least hygienic cat. It is also hardly surprising that sporting this "new, improved" coat, a heavier body build, and a decrease in overall genetic vigor, pedigree Persians are not inclined to patrol territory actively, gambol and frolic, or hunt, and even find sitting on a warm human lap makes them uncomfortably hot. Many Persians are now flesh-and-blood toys, incapable of surviving without human intervention.

Radical genetic mutations occur from time to time in all animal populations. If these are damaging to survival, they quickly drop out of the gene pool. Breeders, however, often cultivate what are extremely rare mutations in order to create novel races—of which the Sphinx and the Rex are two particularly unpleasant examples (see boxes, pages 44 and 45). A normal cat's coat consists of three hair types; the guard hairs, which form the coarse top coat and

The Maine Coon cat, possibly descended from the pampered Angora, maintains its healthy vigor.

The Maine Coon

In contrast to the Persian, the handsome American long-haired Maine Coon has a coat similar to that of non-pedigree longhairs but is a mouser who makes it her business to patrol her farmland territory. The exact provenance of the Maine Coon is unknown but theories abound, one being that when the French Queen Marie Antoinette foresaw her end at the guillotine in the 1790's, she chartered a ship called the *Sally*, owned by one Samuel Clough of Maine, to transport her to the US and evade her doom. The plan failed, and Clough was forced to sail without the Queen, but with all her effects, which included her six much-loved Angoras. Still genetically vigorous at this time, a diet rich in rodent protein, combined with the outdoor life, transformed these once-pampered French darlings into US masters of all they surveyed. Maine Coons also enjoy accompanying their human on patrol.

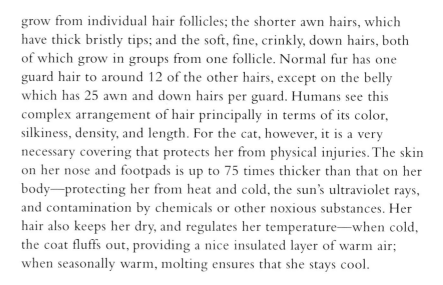

grow from individual hair follicles; the shorter awn hairs, which have thick bristly tips; and the soft, fine, crinkly, down hairs, both of which grow in groups from one follicle. Normal fur has one guard hair to around 12 of the other hairs, except on the belly which has 25 awn and down hairs per guard. Humans see this complex arrangement of hair principally in terms of its color, silkiness, density, and length. For the cat, however, it is a very necessary covering that protects her from physical injuries. The skin on her nose and footpads is up to 75 times thicker than that on her body—protecting her from heat and cold, the sun's ultraviolet rays, and contamination by chemicals or other noxious substances. Her hair also keeps her dry, and regulates her temperature—when cold, the coat fluffs out, providing a nice insulated layer of warm air; when seasonally warm, molting ensures that she stays cool.

The development of compromised designer cats continues apace. The Munchkin, whose tiny stumpy legs prevent her from jumping and climbing; the Scottish Fold, whose deformed, immobile, flattened ears constantly signal a defensive attitude to other cats and which, should it chance to mate with another Scottish Fold, produces offspring with notably deformed cartilage and bone; and the Ragdoll, who goes utterly limp on handling.

Fortunately for the cat, who in Europe alone by the late 1990's numbered over 75 million, the overwhelming majority are either

The Sphinx

The Sphinx, hairless except for an almost invisible down on some parts of her body, is both unable to regulate her own body temperature and acutely vulnerable to injury. Clearly incapable of surviving in anything resembling a normal environment, she must be dressed in woolly coats of varying thickness and confined to living indoors. In North America this unfortunate creature was bred originally from a single hairless kitten born in Oregon.

vigorous moggies or races such as the Korat or Siamese who, not having been selectively bred for the showring, are fit, active, and supremely suited to their geographic and social environment.

Infinitely adaptable, able to survive in arid desert, frozen steppes, luxury apartments, and tropical forests, cats have flourished beyond measure. When their populations could no longer be supported by natural prey, they cleverly took their food from man, increasing their numbers still further.

The canny cat is a masterpiece of design. Selective and incestuous breeding cannot improve her. As every true cat-lover knows, she is perfect already.

The classic moggie-a mongrel cat with a fine gene pool.

The Rex

The Rex appears to be deformed, with curled fur that is retarded in growth, never reaching a normal length. Cornish and German Rexes do not even possess guard hairs, while the Devon Rex is further compromised by having extremely fragile fur that grows from smaller than normal follicles—they are anchored so insecurely that when the cat attempts to groom herself she licks herself bald. All the Rexes are now incapable of adequately regulating their own body heat.

CHAPTER THREE:

TRANSLATING CAT

"The English cat mews, the Indian cat *myaus*,
the Chinese cat says *mio*, the Arabian cat
naoua and the Egyptian cat *mau*. To
illustrate how difficult it is to interpret
the cat's language, her 'mew' is spelled in
31 different ways, five examples being
maeow, me-ow, mieaou, mouw, and murr-raow."
— I.M. Mellon, *The Science and Mystery of the Cat*, 1940

The characterful cat, creature of supreme
individuality, is above all else herself.
Although the Oriental breeds show some
overall differences in disposition,
principally by being more talkative and
friendly—and some work suggests that those
possessing the genetic allele for lustrous
marmalade fur are relatively aggressive and
dominant—*Felis catus*'s massive genetic
diversity ensures real character variety.
Or, to put it another way, when presented
with the same situation, each cat responds
in her own unique style.

Cats in residence at Switzerland's University of Zurich who were presented, one by one, with a human who evinced no interest in their catlike wiles and read a book, took anything between two-and-a-half seconds and two-and-a-half minutes before attempting to engage this deeply boring human's attention. Their behavior style remained consistent over a series of tests, allowing feline researchers Dennis Turner and Claudia Mertens to assert confidently what cat caretakers already know, that cats can "differentiate qualitatively between shy and trustful individuals, initiators of social contact and more reserved cats, and individuals having a preference for body contact or play, and those showing no preference."

Science has proved what cat-lovers already know; that a cat will actively engage you in social activity.

A key factor in human–feline harmony, emphasized by many studies, including Turner's Swiss research, is the acceptance of the cat's intrinsic self—her own unique character, essential feline independence, and love of freedom, coupled with a desire to do exactly as she pleases. As Mertens succinctly put it after exhaustive chez-cat observation: "The more a human behaves in accordance with a cat's activities and preferences for different types of social interaction [e.g. play, bodily contact] the more the cat responds with friendly behavior."

What Can Cats Understand?

The conventional scientific community maintains that only *Homo sapiens* uses and understands language. Increasingly research with other mammals is proving this to be false.

Rico, an intelligent, much-loved pet Border collie, knows the names of 200 different toys and is able to pick them out on command but, crucially, he also makes linguistic inferences. When kids pick up a new word they immediately figure out just what it means. Timmy, who is three years old, is asked to pick out a kangaroo, a creature he has never heard of before, from a group of stuffed animals he knows well. He will automatically pick the kangaroo because he understands that the new name must belong to the new toy. Rico does exactly the same, showing that he understands that objects have specific names and can remember them—even though he can't vocalize them—which indicates that being able to attach meaning to a specific sound [word] evolved earlier than, and independently of, the physiological ability to say it. In other words, while it is only humans who can speak, the development of language is part of mammals' learning and memory mechanisms and, as such, is not the exclusive province of humans.

If the canine race has linguistic ability, then surely so does the equally intelligent, although admittedly entirely differently motivated, domestic cat.

While Turner and Mertens were investigating the mysteries of feline individuality, they discovered that when their subjects wanted to attract human attention they almost all relied on vocalization, particularly when the human's attention was directed elsewhere—which, of course, it generally is in daily life. Conversely humans also learned that by far the most successful way of gaining feline attention was to talk to them, a ploy which frequently induced the cats to rub the humans with their heads or snake around their legs.

How Cats and Kittens Communicate

Kittens and their mothers talk to one another constantly; even from the early age of three weeks kittens reveal a repertoire of quite distinct vocalizations that convey specific messages about their all-important well-being. These vocalizations include: The defensive spit, purring—which allows her to convey her well-being even while she is suckling as it requires her to breathe through her nose while her mouth is closed—plus a variety of distress calls, similar to the adult meow. Calls used to complain of cold are the highest in pitch and fade from use at six weeks old when the kittens can regulate their own body heat. Calls denoting isolation, which for feral or wild cats might mean they had been left behind, are by far the loudest. Calls that alert the mother to the terrible fate of being trapped, which in the litter den can mean their mother is accidentally lying on them, were longest in duration. These "trapped" warnings occur a third more frequently than the "cold" vocalization, and twice as frequently as the call denoting isolation.

Cats are clearly aware that humans have little ability to communicate by scent. Having realized early on that vocalization is an effective means of gaining their mother's attention, cats seem to have understood intelligently that the same ploy will work with humans—to put it another way, they have taught us their language and trained us to respond appropriately to their cries.

By the time they are three weeks old, kittens are communicating with their mother through sound.

The adult cat's main vocalization is the vowel-containing meow, which is capable of multiple variations of sound, emphasis, pace, delivery, quality, volume, and repetition. This makes it ideal for indicating a variety of different requests, complaints, and greetings, as well as for conveying general mood.

The basic meow is a voiced column of air that is modified while the cat breathes in and out. To vary it the cat changes the tension in her mouth and throat, the speed and energy with which the column of air is delivered, and the amount of time her mouth is open, as well as the rate at which she opens it. Vowel sounds are mainly caused by changes in throat tension which lengthens the sound, delivering variations such as meeeeeow, mew, meow, which our cats teach us mean "Food now!" "Let me out," "No," and so on. When alone, cats, just like humans, can often be found murmuring to themselves as they play and go about their daily business, but when a human appears, a cat immediately addresses her vocalizations to them. Are they embarrassed to be caught "talking" to themselves, or merely being sociable?

How Mother Cats Talk to Kittens

Mother cats communicate succinct and important messages to their offspring. Paul Leyhausen, the German naturalist who wrote the classic 1970's cat treatise *Cat Behavior, the Predatory and Social Behavior of Domestic and Wild Cats*, spent an inordinate amount of time studying his own pets' behavior. He soon realized that when the mother brought a delicious mouse home for her kittens, she let them know what was on the menu with a gentle but high-pitched gurgle. When she brought home a rat, or even part of a rat, for dinner she announced her arrival with a scream-like variation on "mouse." The kittens clearly understood the difference between these feline words; they came running in eager anticipation when a mouse was due but were much more cautious when rat was on the menu. Leyhausen concluded this was because a living mouse is no threat to a kitten, whereas a rat that was perhaps simply stunned could inflict severe injury if it came round. It seems, then, these feline communications not only conveyed the image of a specific animal but certain qualities associated with it.

What Meows Mean

Moelk postulated 16 basic feline vocalizations, all of which can be varied, depending on just what your cat wants you to do, or understand.

A] Murmur Patterns, where the mouth is closed and voiced breath passes through the nose:

	Adult	Kitten
1. Purr	1. *hmm*-rmm-*hrn*-thn	1. n/a
2. Request or Greeting	2. mmn	2. *mm*-ing
3. Call	3. *mmn*	3. n/a
4. Acknowledgment or Confirmation	4. *mm*-ing	

B] Vowel Patterns, where the mouth opens and gradually closes and voiced breath passes through the mouth. These are mostly used in cat-human communication:

	Adult	Kitten
1. Demand	1. *mmn*-nn-oow-a	1. *mee*-yow!
2. Begging Demand	2. *mmn*-nn-*aw*-a	2. *mee*-yar
3. Bewilderment	3. *merr*-ra	3. *nng*-a-ou
4. Complaint	4. *ming*-a-(h)ow	
5. Milder forms of the mating cry	5. mmm-oo-(h)ow	
6. Anger wail	6. *woo-oo*-oo	

C] Strained Intensity Patterns, where the mouth is held tensely open and voiced breath is forced through the mouth:

	Adult	Kitten
1. Growl	1. grrrr	1. n/a
2. Snarl	2. *aye*-a!	2. n/a
3. Mating cry	3. ooy-ooy-a	3. *m*-yee!
4. Pain	4. aye-ee!	
5. Refusal	5. *aye-aye-aye*	
6. Spitting	6. *fff*-tu	

How Cats Talk to Humans

Because we hear feline vocalization in much the same way as we do human language—possibly, because like music, it is tonal—we are extremely receptive to its messages. This suggests that relatively complex exchanges may be feasible and hints that real interspecies communication may be possible.

One of the main reasons our cats talk to us is to make us perform tasks for them that they are unable to do for themselves. Unsurprisingly, then, their repertoire of coaxing and demanding vocalizations is relatively large. Even so, when a feline is beside herself with desire for something—perhaps she is on heat and desperate to meet that handsome tom next door and you have neglected to let her out—then, rather like a human abroad, who in a moment of intense passion reverts to his or her mother tongue, she abandons language entirely and resorts to the universally understood, loud, begging wail.

Cats will always try to coax humans with their own language first—before resorting to a begging wail.

Cats also use nonverbal communication to instruct their humans, but will return to the meow if this does not work, proving that verbal communication is the most effective form of understanding between species. Moelk, one of the first researchers to study feline communication, observed that, being confident that you will concede to your cat's wishes, she will place a "heavy stress emphasis on the initial vowel of the vowel pattern." If for some inexplicable reason you still fail to comply with your feline's wishes, she will lapse into complaint mode, which stresses the first syllable and becomes ever louder. Further stubbornness on your behalf gives a jolt to your cat's self-confidence, resulting in a "weakening of the initial vowel stress, the lengthening of the final vowel and the raising of the prolonged final vowel in bewilderment."

Cats frequently feel when things are not entirely as they should be, causing Moelk to observe that "almost any unoccupied cat who is not in a state of maximal comfort … as represented by drowsing upon a full stomach in a warm, soft, safe spot with friendly hominids [humans] on hand, is in a sufficiently uncomfortable and disturbed state to find ready release through the noises of complaint," and naturally feline vocalizations reflect this.

Following up this "complaint" theory, Moelk took to the streets greeting every strange feline she encountered with a friendly interrogative greeting, such as: "What's kitty doing there, hmm?" Out of 12 cats, none greeted her—cats normally acknowledge familiar or friendly humans—while nine actually signaled feline discontent. Moelk also discovered that her interrogative tone was particularly effective in inducing further affectionate responses, such as purring or head rubbing. For an interpretation of What Meows Mean see pages 52–53.

Interspecies Communication?

Although it is principally the cat who teaches you her language you can also school her to recognize a whole variety of human words and phrases—from her own name to "roast chicken for supper." If you use the same words and, crucially, also the same intonations every time you take either chicken or salmon respectively out of the refrigerator and devotedly put it in her bowl, your feline will soon understand from your vocalizations alone whether it's chicken or salmon for supper—like the kittens who quickly learned whether rat or mouse was on the menu and behaved accordingly.

Once your cat can differentiate between these and other food-item vocalizations you should find it possible to ask your cat what she wants for supper as she will register her desire for a particular item by replying with a demand call. Using the same methodology her vocabulary can be extended further to include words for her toys, for catnip, or simple desires such as "do you want to go out?" But can your cat understand complex sentences? Will you ever be able genuinely to converse with her? True interspecies communication has always been looked on as an impossibility by the conventional scientific community because it denies that animals possess language or consciousness. More realistic objections are that besides vocalizations animals rely on a variety of complex means of communication, such as smell and scent, which we are not equiped to understand, and in most species, their physical make-up inhibits them from making the same type and variety of sounds as we do.

However, recent work with African gray parrots, who can make the same vocalizations as us, and who frequently attain vocabularies of over 500 words, is beginning to prove otherwise. Aimee Morgana is not dedicating her life to teaching her African gray, N'kisi, complex tricks or testing his cognitive ability, but to discovering what he is actually thinking. To this end she has spent more than four years teaching N'kisi to converse with her. The result of all this effort is that N'kisi knows more than 700 words, formulates his own sentences and often comments about what's happening around him. As far as I am aware, to date, no one has applied Morgana's approach to a feline, yet *Felis catus* has developed a vocabulary specifically to speak with us, and has a tonal language containing vowels (see pages 52–53) So despite certain morphological limitations on some consonant pronunciation, the domestic cat seems to have potential for the development of interspecies communication.

Understand Your Cat's Own Words

Record your feline's calls and—with the advent of tiny digital tape recorders which are easy to position at cat level and whose contents can easily be downloaded—analyze them. Talk to her, and record her replies and her vocalizations when friends arrive or a cat walks by, and keep a detailed log of everything that is happening when she "talks." By replaying the recordings and viewing your log, you will uncover repetitive vocal patterns linked to activities, demands, perhaps even states of mind, and you may discover your cat is telling you far more than you could ever have imagined.

The Language of Smell

Although cats principally use sound to communicate with us, in their world smell is of vital importance. Through scent-marking cats leave detailed messages both for neighbors and strangers, warning that a territory is already taken, and advising of their status and even their sexual availability. This "scent" newsletter prevents cats straying accidentally onto one another's home ground; it also reduces the incidence of armed combat and may even lead male cats to romance.

A scratching post is both an essential feline notice board and a useful claw-sharpening facility.

Scratching posts are both an essential feline notice board and a claw-sharpening facility. Claws need to be maintained in premium condition and dragging them down the bark of trees, which flakes off slivers of keratin—the materials our nails and their claws are made of—keeps them admirably sharp. The act of scratching also activates the scent glands between the cat's toes and this, combined with the information left by her paw pads' watery sweat glands and the very visible scratches, make the ensuing message a potent and complex one. Scratching posts are usually sited on much-used trails, which would seem to indicate they are not territorial markers but purely sources of information.

Inside your home the cat will favor soft wooden surfaces—pine table legs are ideal—or padded upholstery. Once a cat has established a scratching post, she uses it continually to emphasize her message. It is possible to prevent this behavior indoors by removing all traces of the scratch marks as soon as they occur and dousing the area with something extremely strong smelling. Crushed mothballs are ideal.

If she persists put a commercial scratching post adjacent to her favored new spot, and rub a few drops of essential oil of valerian on to it. Your cat is unlikely to need any further encouragement to

use this new post as cats find valerian unbelievably delightfully intoxicating. Although at first you may need to transfer her to it, one whiff and she should soon abandon your table legs. Playing with the post with her, perhaps by dragging her toys over it, is an added inducement.

The glands between the toes aren't the cat's only scent glands, she also has them under her chin (submandibular), at the corners of her mouth (perioral), on either side of her forehead (temporal), at the base of her tail, and distributed evenly along it (caudal)—even the grease of her coat leaves a potent message. She leaves her own perfume on key points, such as those that have been marked by another cat or that possess her caretaker's smell, by rubbing them with her head. Your cat can also smell if an intruder has dared pass through her catflap.

Cats who live in groups, and in contemporary times this is increasingly usual, be they feral or domestic, frequently rub their faces together, creating a kind of flexible group identity, a tribal marking, which gives confidence and assurance to its members. Our felines regard us as part of their group, and rubbing their heads on our legs, and encircling us with their svelte bodies, allows them to not only give us their perfume, and that of their feline group, but receive ours and include us in their tribe. Confident felines will leap onto the sofa to exchange head perfume, although some find our giant flat faces rather intimidating. If we are not available, our abandoned jumpers and warm duvets give them both comfort and the ability to indulge in further mutual perfume spreading and to strengthen ties of affection and friendship. If you have been dallying with a feline from an alien group, expect to be greeted with a hiss instead of your feline's normal welcome call.

Body Language

Facial Expressions

Cats also employ facial expressions, ear movements, body posture, and eye contact to signal a range of different emotions, intentions, and moods. Some of their most powerful messages are conveyed by gaze. Except in the most passionate of courtships, staring is an aggressive threat and is taken as such, whether the starer is cat or human. When guarding her own territory, a cat will pose sphinx-like, for literally hours at a time, staring intimidatingly at a cat who has transgressed the borders of her catdom, who returns like with like until, finally, one retreats or full-scale hostilities commence.

The combination of an even gaze and relaxed tail position indicates two cats happy in each other's company.

In the latter situation the aggressor's pupils will be narrow and focused while those of the defensive feline will be open wide. When not engaged in dispute, enlarged pupils indicate an alert interest. If your cat finds her new toy pleasing or her latest delicacy acceptable her pupils will dilate; if not, she will no doubt let you know with a complaint call. Half-closed eyes and slow blinking are signals of feline reassurance. As we possess no tail with which to signal our friendly approach, combining these with an averted gaze, an interrogative tone of murmurings, and dropping down to cat level and letting her approach you, are extremely useful tools to use when trying to make contact with a strange or nervous cat. Cats use their facial expressions and body language to convey emotions and intentions to one another, in ways which are so subtle humankind is yet to understand them. As body language is not their principal mode of communication with us, it does not compromise the feline-human relationship. However, the understanding of classic defensive and offensive postures will help you gauge your feline's mood.

Facial Expressions

This chart, based on the illustration in Leyhausen's classic work, *Cat Behaviour, the Predatory and Social Behaviour of Domestic and Wild Cats*, shows a feline face morphing through various degrees of defensive and offensive mood.

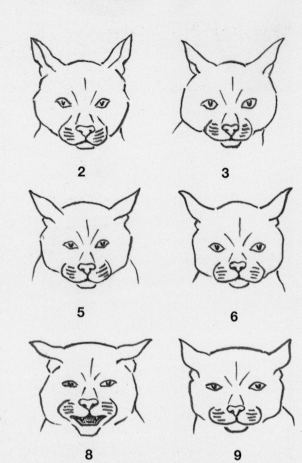

1   2   3

4   5   6

7   8   9

Top left (1) shows the feline in everyday relaxed mood; the top line shows her becoming increasingly aggressive until she is on the verge of attack (3), indicated by the amount of the back of her ear that is visible. If we move down from the top left expression to the bottom left (7) we see her becoming increasingly defensive and determined to defend herself vigorously. This is indicated by her ears, ever-increasingly folding downward and sidewise. Other central and bottom row squares represent a mingling of these emotions, until we see in the bottom right illustration (9) a cat who is indecisive but ready to adopt whatever offensive or defensive strategy necessary.

Ear position is crucial in determining a cat's mood; and it can change in an instant. Because of this, many cat breeds, such as the glorious Maine Coon, have tufts of hair on the tips of their ears, making their position more readily visible, even from a distance. According to Leyhausen the greater the amount of the back of the ear that is visible from the front, the more determinedly aggressive the feline is feeling. An intensely defensive feline's ear is folded sidewise and downward, allowing only the merest slither of the inside of the ear to be visible and, if possible, it is completely concealed. Ears are vulnerable to teeth and razor-sharp claws in cat fights (as the tattered ears of many a bold tom cat attest), folding them down helps to protect them. The moment an aggressor strikes, a cat's ears flick flat, removing that vulnerability.

Expressing Discontent

Cats use vocalizations to express their exhaustion, pain, and unhappiness. Surprisingly, cats can purr when in chronic pain, physically traumatized, debilitated, or terminally ill. It may be that purring comforts and calms them as it once did when they were kittens, serving the same purpose as rocking from side to side does for humans, as well as communicating the self-protective message: "I am only small, helpless, inoffensive, and innocuous." If your cat seems even slightly off-color and is purring constantly, it is always worth taking her to the vet.

The Cat Thermometer

The degree of your feline's body curl position reveals her temperature. Cats are very sensitive to heat and make every effort to maintain a steady, even temperature and a relatively low metabolic rate. With no sweat glands to help her cool down, a cat will move from sun to shade to adjust her body heat, while in winter she will puff up her fur to keep a layer of warm air next to her skin.

Cat thermometer based on a study of 400 positions by the German animal expert Dr Hans Precht (from Incredible Cats *by David Greene).*

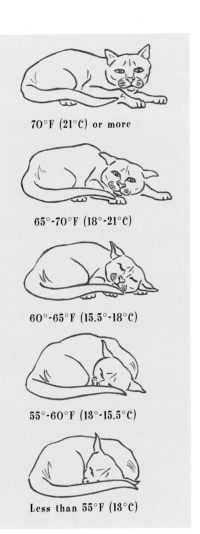

70°F (21°C) or more

65°-70°F (18°-21°C)

60°-65°F (15.5°-18°C)

55°-60°F (13°-15.5°C)

Less than 55°F (13°C)

Why Cats Don't Submit

When a dog lies on her back she is either showing ultimate trust or groveling to the greatest degree; when a cat takes this position it is the ultimate aggressive defense, the optimum position for inflicting damage on her opponent's soft underbelly or vulnerable throat with her razor-sharp claws and blade-like teeth. When your cat adopts this position on the sofa she is certainly not signaling that she considers you her superior, although when she allows you to stroke her belly, she is showing you deep trust and affection. But even so it is a rare feline that accepts this caress for more than a moment. Her discomfort can be gauged by the twitch of a tail, which, if ignored, becomes more agitated until it escalates into a thumping movement. If she has not already swatted your hand with her paw, be prepared for that and more. The moment your cat issues the first tail twitch is the moment to desist. This simple act may even calm her sufficiently to let you stroke her again. However if you persist she may fasten onto your hands with her claws or even sink her teeth into your flesh. If this happens don't try to wrench your hand away—she will bite harder. Instead freeze—this reduces provocation, and calms her, inducing her to let go.

Your cat can, with patience, be taught to recognize vocal patterns representing "fish," "biscuits," or "dried pellets."

Body Language

As in the Facial Expressions chart on page 60, the body postures below illustrate the morphing of a cat's mood from offensive to defensive. However, body positions are less reliable indicators of intent than a cat's face, which changes with lightning speed. The body may lag behind in shades intensity or direction.

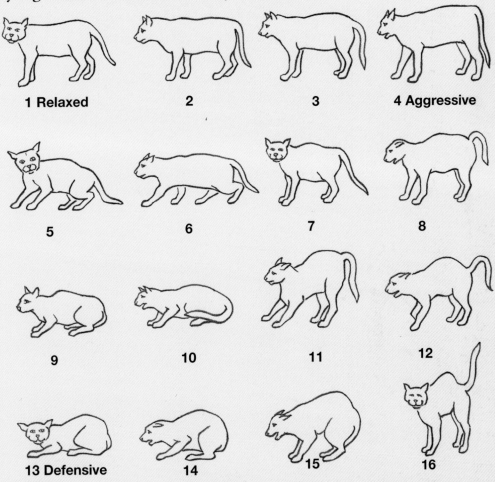

1 Relaxed　　**2**　　**3**　　**4 Aggressive**

5　　**6**　　**7**　　**8**

9　　**10**　　**11**　　**12**

13 Defensive　　**14**　　**15**　　**16**

The loser in a cat fight, be it a sphinx-like staring match or a grueling physical fight, will immediately give up the territory and leave. For this reason open submission is not an emotion felines are familiar with. Adopting their most defensive body posture, bottom left, can sometimes cut off attack but this is probably because the aggressor realizes just how ruthlessly this feline is prepared to defend herself and may consider the prize not worth the risk. It does not in itself carry any admission of defeat.

The Evolution of "Tail Up:" The Cat Greeting

Felis catus, and all her wild relatives from the mighty tiger to the tiny Sand cat raise their tails vertically—Tail Up—when spraying, and lower it immediately afterward. But the upright tail is now known to be a sign of greeting and inclusion given by the domestic cat.

As observation of many members of the *Felidae*, and in particular *Felis silvestris*, has been sparse due to these creatures' illusive nature, Cameron-Beaumont at the University of Southampton decided to observe one member from each of the three *Felidae* lineages—the caracal (panthera), the Jungle cat (domestic), Geffroy's cat (ocelot)—to discover if they used Tail Up in any other context. Five-hundred-and-thirty-nine observation hours later, it seemed they did not. What literature on felid communication there is, and it is sparse, concurs on all species except two: The domestic cat, and that other most social of felids, the lion.

Panthera Leo, the lion, was first observed indulging in greeting and social inclusion rites using an upright tail by the great naturalist and conservationist George Schaller in the Serengeti. Tantalizingly he wrote, "During head-rubbing and anal sniffing contacts the animals raise their tail so that it either arches over their back or tips toward the other other animal," but gave no further contextual information. However more recently Estes has noted that the Serengeti lion-pride members have a "common greeting ritual of rubbing heads together with tails looped in the air, while moaning."

So why should these two species have developed Tail Up as a greeting rite? Probably because lions, like domestic cats, live at higher population densities than other feline species which has led to their forming social groups, whose territory they defend against other groups and whose membership they defend against alien loners.

Group members generally lift their tails upright when approaching for rubbing or other scent exchanges and if the object of their interest is in accord they too, will put their Tail Up. Indeed, intensive observation of a colony of neutered cats at Southampton University by Cameron-Beaumont revealed that tails were up more than 80 per cent of the time during rubbing and sniffing rituals.

Tail Up also has the advantage of being a highly visible and unmistakable signal and *Felis catus* will sometimes keep her Tail Up for prolonged periods of time. This means a colony member returning from a lone excursion can easily signal her group connection and friendly intent, while a lone cat, perhaps seeking affiliation, can also show that she comes in peace. This may mean the latter can be ignored instead of attacked, greatly conserving energy. As cats are heavily armed in tooth and claw, Tail Up would seem to be a useful evolutionary device. The naturalist Roger Tabor, who closely followed a colony of cats who lived in Fitzroy Square, west London, during the 1970's and 1980's, observed the cats adopted Tail Up to one another as they massed to meet their self appointed human feeder Mary, with her laden trolley, and greeted Mary in the same way. However the colony's tails remained firmly down when non-members approached.

As you are a member of her group, your cat greets you and other humans she may be interested in with Tail Up and quite often a friendly call, for a variety of reasons ranging from "hello," through wishing to be caressed to hearing you open the refrigerator and desiring a snack. And every time you return this salutation by stroking her neck or her back, you confirm that you understand her, that you really do speak cat. But Tail Up isn't just about signaling friendship or affirming group membership; it can also be an invitation to play or for you to play tag. Jeffrey Masson, who spent a year observing his five cat companions in order to write *The Nine Emotional Lives of Cats* reveals: "The moment he [Miki, one of the five] sees me his tail goes straight up. At first I did not recognize this for what it is— a sign of friendship mixed with confidence, but now I have seen it, I see it everywhere. Happy cats lift their tails up, and judging by how they modify their behavior, immediately responding to the elevated mood as if happiness were infectious, we know cats understand this."

Tail Talk: What Your Cat's Tail Is Trying To Tell You

1. Tail at angle of 15–20 degrees from legs: Relaxed. •

2. Tail lifted and curved slightly over her back: A general expression of everyday well-being and confidence. The cat may move her tail in time to the rhythm of her walk.

3. Tail Up (see pages 64–65) : An initial friendly greeting and/or a signal of interest, expectation, and demand; it is also frequently seen as an invitation to play and an expression of a cat's general happiness.

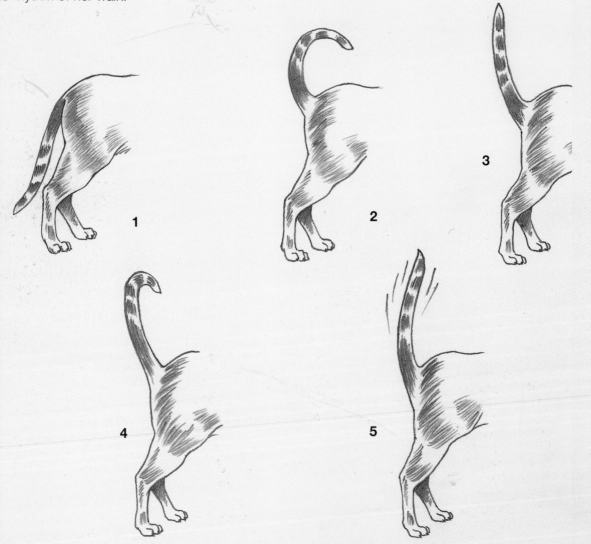

4. Tail Up but the end of the tail bent over like a half question mark: Greeting.

5. Tail straight up and quivering: The spraying position and, some say, when directed at us, an expression of ultimate affection.

6. Tail held straight down: Indicative of imminent, although not yet irreversible attack.

7. Tail flicking back and forth: A sign of irritation, or annoyance.

8. Tail lashing from side to side: A sign of intense anger and a warning that an attack is imminent when combined with other aggression indicators (see pages 60 and 63).

9. Tail arched and down: Aggressive state of mind.

10. Tail fluffed up and arched: Combined with an arched back, this feline is torn between aggression and defense, but as the tail reaches the horizontal aggression increases and defense decreases.

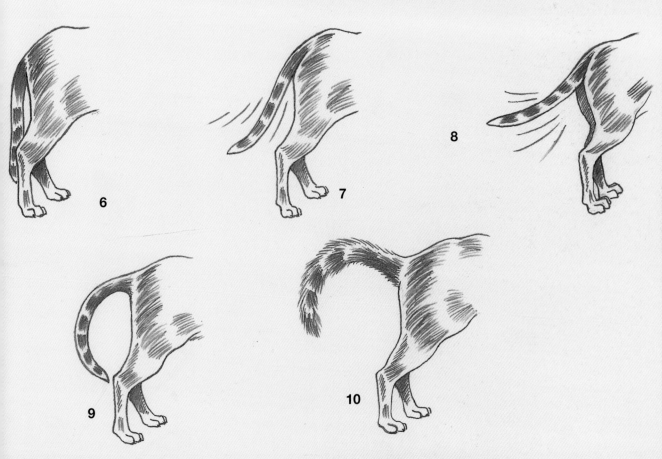

Understanding Problem Behavior: What Your Cat Is Trying to Tell You

Since the 1980's changes in human living and working patterns have meant larger numbers of felines are being confined indoors for long periods and it is no coincidence that reports of problem cat behavior have rocketed. This is partly because humans who contain their cats at home may expect them to be far more affectionate and predictable in nature than cats that are allowed to wander wild and free.

At the same time hand-in-hand with confinement has come a rise in multi-cat households, as people believe that company for their indoor cats alleviates boredom; although in many cases this can lead to stress as a result of the cat's natural territorial imperative.

Indoor Spraying in Cats

Cats confined inside the home can feel restricted, and relationships between pets may also cause stress.

Contrary to popular belief, both male and, more rarely, female cats can spray. Illnesses such as feline urologic syndrome—two major components of which are cystitis and urethritis—can cause both indoor spraying and indoor urination. While the first port of call should always be to the vet to rule out these conditions, spraying is usually caused by stress. Outdoor cats come and go as they please, while indoor cats are denied basic pleasures. Instead they remain home alone in what can be a monotonous environment, which can lead to behavioral problems. Confinement is seriously stressful for felines, even *Felis catus*. Lacking the rich variety of stimuli and experiences that millions of years of evolution have equipped them for, they can become bored and unable to express themselves. Denied access to an outdoor territory, which cats will mark by spraying on vertical message posts such as trees, telegraph poles, and rocks, the indoor cat instead marks the only territory available to them in their luxury cell, by urinating on drapes, loudspeakers, and sofas. Of course, even outdoor cats spray indoor objects occasionally—typically your baggage when you return from vacation, or that luxurious new couch. This is not, however, some form of terrible feline vengeance or jealousy; your cat merely wants to make these objects hers, to include them in your mutual territorial zone. Left to their own devices in the wild, cats can also leave feces on frequently used trails and at junctions. With no external access, a hallway becomes a trail. This ingrained behavior can be intensified by additional stresses, such as a rival feline strutting past the window, or noise, dogs, visitors, and the addition of a baby or another cat to the household. It escalates further in spring, when hormonal changes also incite cats to mate.

A number of drug-free behavioral solutions can help calm your cat and thus induce her to stop spraying. Most importantly, don't punish her; the added stress of human conflict may even worsen the problem. Choosing to neuter or spay the cat should reduce the spraying, particularly if your cat is male. If you keep an indoor cat, whom you suspect may be feeling threatened by cats from outside, invest in a harmless motion-activated device for the yard that discourages feline intruders. If neighborhood cats are taunting her from outside, just close the drapes. Alleviate boredom with distracting toys that need not be expensive—tie a small ball to a long piece of yarn (not nylon, which can harm cats) and leave it dangling from a door handle, or make a shaker toy with a handful of walnuts and an old jar so your cat can amuse herself if she is home alone. When cleaning up a deposit, use enzyme-, not ammonia-based cleansers, as ammonia may smell like urine to your cat and provoke another episode. Covering favorite spray spots with plastic wrap will both deter the cat and protect the area. If the spraying results from conflict between resident cats, ask advice from your petcare professional. Finally, research shows us that all cats who sleep in bed with their humans become more settled and confident.

If your cat is being stressed by taunts from neighborhood invaders, simply close the drapes.

CHAPTER FOUR:

THE CAT'S OWN AGENDA

"The only really happy cat is a free cat, one that is able to wander out at will, to climb trees or to step soundlessly through long grass, to hunt mice and rats, to sunbathe on roofs, to seek the solitude that its nature demands."

—Jeremy Angel, quoted in *The Nine Emotional Lives of Cats*

The cat is so alive. Every fiber of her body tingles with sensation, with the sheer joy of being, whether she is rolling languorously looking for love or dispatching a rat. Charismatic, charming, and clever, with an unsurpassed individuality, the cat is both the most fascinating of the creatures that lives with us, and the most infuriating. Yet it is the cat's supreme independence, her seeming insouciance-in the face of anger and lightly smoked salmon alike-that make her so irresistible and, ultimately, so forgivable.

Who Knows the Heart of a Cat?

Certainly not the scientists who, even in 2002, asked, "Do cats think?" only to state risibly: "As yet the astonishing differences between individual cats are largely unexplained in terms of both how they are generated and *why they might exist*."

Character is created by the constantly moving interface between genetic inheritance and personal experience, both of which in *Felis catus*, are as diverse as, if different from, humankind. Variety bestows enormous survival advantages on a species as a whole. It may be the one individual who does things differently who holds the key to survival in a changing world. Genetic diversity also allows the species to survive in even the toughest of circumstances. In Chicago at the turn of the nineteenth century cats were put in cold storage warehouses to kill rats. Around 80 per cent succumbed to their grim conditions; 20 per cent survived and bred. In a few years all the warehouse cats had thick warm fur and much shorter tails (extremities always suffer from cold and frostbite first).

The brown Burmese; vacant or mysteriously deep in thought?

Mainstream science has a deep-rooted problem when it comes to crediting animals with sentience and forethought, an attitude that over centuries has seamlessly become part of Western thought. Partly this is a legacy from Christian doctrine, which has denied that animals have souls and stated unequivocally that God created animals for humankind's use. It's also partly a legacy from the Enlightenment of the eighteenth century, which denied the existence of any phenomena if they could not be explained by physics or chemistry. This allowed Descartes' mechanistic view of animals as furry automata who neither suffered nor thought to gain ascendancy throughout the scientific and, later, the wider world—a convenient concept for scientists to hang on to, as it allows global companies to conduct experiments on animals with what they consider to be a clear conscience.

How do Cats Think?

The mainstream scientific community has labored long and hard to maintain the fiction that animals lack sentience, self-consciousness, and the capacity for abstract thought. To this end it pours scorn on anecdotal evidence and even careful observation by naturalists, claiming them to be unscientific and unverifiable. In turn science conducts experiments on animals that require them to perform a task containing a peculiarly human characteristic such as admiring themselves in a mirror.

In 1997 a feline named Bill became an object of research. Bill pottered around the lab in the day, as well as having his own room, which was attached to an extraordinary feeding apparatus. This contained moist cat food on which he could dine to his heart's desire, but to get the food it he had to press a lever. Sometimes he had to press the lever 40 times, sometimes thousands (the research paper does not reveal if Bill suffered from repetitive strain injury). Schedules that varied the number of times Bill had to press the lever carried on for 60 days at a time; schedules where the number was constant—the maximum was 2,560—were ten days long. The variable schedule revealed that Bill averaged out the amount of food he ate over time, regardless of how many times he had to press the lever. The fixed schedule showed that the lower the number of presses, the more small meals he ate; the higher, the fewer larger meals he ate.

Does this extraordinarily contrived experiment reveal anything about Bill—a perfectly designed predator, not a perfectly designed bar presser—about his essential catness, or how he thinks? Did Bill despair when after pressing the lever 7,000 times there was still no food available? Did succulent fresh mice populate his dreams? Did his shoulder muscles ache? Did he hope against hope that the researchers might give him a treat?

The mainstream scientific community doesn't yet acknowledge that cats think, let alone understand how their minds work.

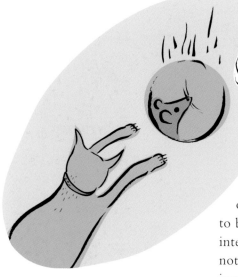

Placing a piece of stolen beef by a rat hole, the cat waited silently before pouncing when the rat was tempted out.

Contrast that information to that given about a tomcat belonging to a man named Callendar who, back in the days when servants were more usual, was seen "bearing away a piece of beef in his jaws. The servant followed and watched him lay the morsel down near a rat hole. Then he hid himself. Presently the rat came out and was dragging away the meat when the cat pounced on him."

Scientists would argue that perhaps the cat dropped the meat by accident; that it was a coincidence the rat came out of its hole and the cat killed it; that the experiment would have to be repeated in the laboratory many times to be certain of feline intent. But cats are strategic opportunists who do as they please, not as we please—and these kind of situations are difficult, if not impossible, to replicate.

A naturalist might say this anecdote tells us that the cat is crafty, that he is capable of forethought, that he remembers where the rat lives and knows what tempts him. More than that it reveals he has what devotees of Piaget's psychological theories dub "object permanence," that is, the mental knowledge that something continues to exist even when it has been hidden from view—in this case the rat in its hole.

When prey hides from *Felis catus*, the cat understands perfectly it still exists and "worries" it out of its lair. Scientists felt the need to prove this. To this end they taught several cats that if they touched a toy with their nose they would get a treat. This toy was then put in an empty cup in front of the cat and placed behind one of two screens. The researcher then emerged with the empty cup and showed it to the cat. Not surprisingly on virtually every occasion the cat went behind the correct screen to retrieve the toy.

Further so-called evidence that most non-human animals are not self aware comes from their seeming inability to pass Gordon Gallup's mirror test—its premise being that only creatures who can recognize themselves in a mirror are self aware. Working from the State University of New York, Gallup put sociable chimpanzees in solitary confinement with nothing for company but a full-length mirror, for what must surely have been an interminable ten days. Initially the chimpanzees tried to socialize with their mirror image but after a few days started to use the mirror to groom themselves, showing that they recognized their own image. To prove his point Gallup then gave the chimps a general anesthetic and painted red marks on their foreheads. When they came round they gave no sign that they were aware of the mark, but when put in a cage with a mirror became extremely curious and touched the strange mark constantly.

Cats, apparently, roundly fail Gallup's test and certainly Leyhausen, the German feline scientific behaviorist, reported that once his cats had established that the mirror image was just that—an image—they immediately lost interest in it.

However, Bingley, a naturalist writing in 1813, saw things differently:

Genetics shows us that just one cat doing things differently can hold the key to success in a changing world.

"No experiment can be more beautiful than that of setting a kitten, for the first time, before a looking-glass. The animal appears surprised and pleased with its resemblance and makes several attempts at touching its new acquaintance; and, at length, finding its efforts fruitless, it looks behind the glass. It then becomes more accurate in its observations; and begins, as it were, to make experiments by stretching out its paw in different directions; and when it finds these motions are answered in every respect by the figure in the glass, it seems, at length, to be convinced of the real nature of the image."

The Mirror Test

Why not test your own cat? Place her in front of a mirror large enough to reflect both her image and yours for a few minutes every day. Notice if she catches your eye in the mirror, something dogs, who also fail Gallup's test, often do—which must mean they understand that it is their image in the mirror, just as it is their companion's. After a couple of weeks, apply some very bright non-toxic children's play paint—cats do not have very good color vision—to her forehead while she is sleeping off her dinner and when she awakes, pop her in front of the glass. If she notices the paint (for example, by trying to rub it off with her paws) then your cat is a certified sensate genius.

Are Cats Self Aware?

Recognition of one's own image may prove self-consciousness, but does lack of recognition, if that is indeed what it is, prove lack of self-awareness and thus, as far as scientists are concerned, inability to empathize, to understand others' points of view, to be aware?

Cats groom themselves constantly, not only to maintain their coat in premium condition—the spines on their tongue's surface act like a comb—but to regulate their body heat. In hot conditions their tongue acts as sponge, spreading moisture all over their body, which then evaporates, reducing their temperature by up to a third. Conversely, when cold, cats use their tongue as a brush to help maintain body heat by trapping air in their now nicely fluffed-up coat. Grooming also spreads their perfume evenly over their body and re-establishes their olfactory identity after handling by humans.

What evolutionary purpose would a cat grooming herself in a mirror serve? Why should she be interested in what she looks like? Is she going to style her fur in a new and fashionable way?

Humans are particularly interested in, if not obsessed with, the visual, and their own appearance in particular, which, due to the different balance of their own sensory system, many other sentient creatures are not. Science constantly mocks the anthropomorphic but what could be more anthropomorphic than expecting a cat to spend time gazing at herself in a mirror?

Perhaps it would be more pertinent to discover if a cat recognizes her own smell.

The Thrill of the Hunt

Much of the pet cat's ingenuity is directed toward one aim—obtaining food, be that superior quality provisions that her human caretaker is selfishly keeping for herself, or flesh from the world's abundant pantry. A rat cannot be dispatched with the same insouciance as a mouse, even by such a supreme predator as *Felis catus*.

The rat is dangerous opposition–it must be stunned into submission before delivery of the coup de grace.

The rat is a dangerous opponent, often almost as large as the cat. It will rear on its back legs, ready to deliver ripping bites with its dangerous sharp teeth, and even a superficial flesh wound can cause infection and ultimately, for some, death. A rat must be stunned, battered into submission with hard paws and claws, before a cat can deliver the coup de grace. Roger Tabor watched an urban rat up against a wall being confronted head on by a cat. The rat threatened in turn; the cat, taking this seriously, backed off, then quickly span on her haunches so she was parallel with the rat and punched its head with her paws.

"It was now the rat's turn to spin around and it tried to go back rapidly the way it had come, giving voice to squeaks as it went. The cat meanwhile beat it until the rat had moved from the range of its paws, then coming down on to four feet it took a few paces to come alongside the rat and beat it with its paws again. It did this until the rat had almost reached its escape home, when the rat perhaps fearing this escape was being cut off by the cat, returned to the original route, pursued with beatings by the cat. The change of route along the wall was repeated once more by the rat, but it was not able to shake the cat."

Mousers, loving the hunt, kill far more mice than they will ever eat, which is of course why they have been so invaluable to man—keeping his granaries free from greedy rodents, dispatching vermin in the trenches of World War One, and more prosaically, in the early twentieth century, protecting the seed supplies of the feathered denizens of the pet department of large general stores from depredation, while carefully refraining from harassing the birds.

Some cats use excess carcasses to stock larders in case of shortage, and in the following anecdote one apparently extremely sagacious feline went even further, providing himself with a convenient living larder.

Killing far more mice than they can eat has made cats invaluable in granaries and larders for centuries.

"Émile Achard's Matapon, having killed all the mice in the house, took to killing field mice. This was difficult and unpleasant on rainy days but it was not long before he conceived the idea, and carried it out, of restocking the house. He brought field mice in alive and let them loose, thus establishing a new hunting preserve."

Others Bring Their Prey Home to Their Humans

A friend of mine who owned two beautiful black-and-white cats lived next to a canal teeming with life. Acting as one unit the cats would smartly dispatch mallard ducks, together drag them up a ladder to their cat flap and deposit them centrally on the living room floor. Very dead, cold carp were always placed lovingly on my friend's pillow while he was out, and a selection of rodents were often to be found on the sofa or sometimes near the refrigerator.

Another friend has three free-ranging cats; when one brings home a dead mouse, the other two are invited to inspect it, after which it is presented to their caretaker.

Why do cats do this? Are they showing off—letting us know just how fabulous at hunting they really are? Is it a feline tribute, a thank you for all the cream and poached chicken, and the warm bed and comfortable velvet cushions? Most naturalists and behaviorists speculate that the cats, feeling maternal, bring home food to what they perceive are useless hunters. But surely the fact that it is we who constantly provide them with food, effectively mitigates against this theory? Others say the assorted small rodents are brought chez-cat to prevent them escaping should they be merely stunned instead of dead. But as felines kill far more animals than they need—some mousers can be seen with literally mouthfuls—this seems an unnecessary precaution, although perhaps explains some cases.

It does however seem perfectly logical to assume that when domestic felines put what is, to them, the most valuable and important thing in their world on their caretaker's pillow, or favorite chair, sofa, or cushion, it is a token of their affection and regard—and the only thing that is theirs to give.

Acting as one unit two cats would place gifts of love; very dead, cold carp on their owner's pillow.

Ingenuity in the Home

But let's consider feline ingenuity in their home zone. This is often directed to opening refrigerator doors and in the past when chillers had large external handles, this was a cinch. In contemporary times smooth unyielding surfaces, and doors heavily laden with bottles of cooling Chardonnay can prove more of a problem—but not an insurmountable one.

The cat of the English actor and comedian Alan Davies soon discovered that by lying on his back, fitting his prehensile paws in the ridge between door and refrigerator body, and bracing himself, he had sufficient strength to lever the door open. On one memorable occasion he was found making good his escape through the cat flap with a turkey, which he proceeded to devour at top speed until it was finally wrenched from his paws.

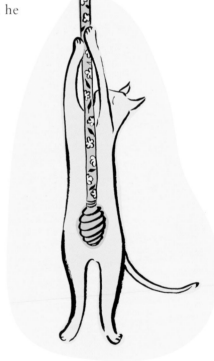

In Victorian times when long bell pulls dangled in every room, cats soon learned not only to summon humans to let them in and out, but to distract hapless servants, leaving the cats free to plunder delicious, now unguarded dinner trays. And, of course, besides using their paws to shade their faces from the sun and the fire, they use them to scoop water and cream from jugs.

In Victorian times cats soon learned to summon humans using the long bell pulls that hung in every room.

Obscure Objects of Desire

Cats' sensual desires don't stop at food and sex—they are also devotees of intoxicants. Some lick beer from propitiously leaking beer taps until drunk, while others favor mulled port and rum punch. As to a taste for the latter, cat's companion Miss Savage refreshingly and unsanctimoniously noted in a letter to Victorian novelist Samuel Butler, dated 24 December 1879: "Poor old dear, he is all the better for it."

For the estimated 50 per cent of cats whom, thanks to their genetic inheritance, can enjoy its narcotic delights, catnip, or *Nepeta cataria* to give it its Latin name, is the intoxicant of choice. Most cats sniff then lick or eat catnip, but those who feel its effects keenly, pull the herb to their head and rub against it. In this state of ecstasy felines will strike, claws unsheathed, at anyone who attempts to deprive them of this delight. The molecules of nepetalactone, the active ingredient of catnip, follow the same biochemical pathways through the brain as those of marijuana and LSD, and seem to induce the same variety of psychedelic states humans experience—causing cats to lapse into a trance, stare meditatively into the vast infinity of space, chase phantom mice, or become extremely aroused, and clearly cats enjoy this as much as many humans do.

Catnip Magic

The cat has been of supreme importance as an amulet and charm throughout history. To attract *Felis catus*, those who practiced the subtle arts—black and white—for love or hate, for wealth or pain, played on her love of catnip: "In the new moon gather the herb Nepe and dry it in the heat of the sun; gather vervain in the hour of 8, and only expose it to the air while the sun is under the earth. Hang these together in a net in a convenient place, and when one of them has scented it her cry will soon call those about within hearing; and they will rant and run about capering to get at the net, which must be hung or placed so they can not accomplish it, or they will tear it pieces."
Conjurors Magazine, 1791
According to M. O. Howey, author of the seminal work *Cat in the Mysteries of Religion and Magic*, there is a place known as "the field of cats" near the city of Bristol in the UK, which is named after the huge number of cats attracted there by a conjuror's spell.

Tribal Living

Felis catus's real nature belies her image as "the cat who walks alone." Almost all cats live in groups or tribes that have their own unique structures and, like humankind, much of this societal form depends on the physical space available and the abundance of food. In the past the density of people in rural subsistence farming in India worked out at one per two acres (one hectare), which provided sufficient produce for survival. But, of course, the people clustered together in groups, to use cat terminology, shared a core area, and farmed home ranges which sometimes overlapped.

In rural England, for exactly the same reasons, the density of feral cats is around one per 20–25 acres (8–9ha), and like the Indian farmers, they cluster together in small tribes, socializing in a core area from which they meander out onto their home ranges. However, the female cats have home ranges that cluster together, while male ranges overlap them and are on average ten times their size. Frisky toms mate with females in their own range but will court females from other groups if they can get away with it.

In cities human density can afford to be immense because food is shipped in and distributed through a complex network—and the same applies to cats. In central London, where "railings provide shelter, scavengeable food is accessible and generous auxiliary feeding is provided by a battalion of feeding ladies," who act as a kind of feline superstore—cat density is about one per 0.2 of an acre (0.08ha), with a social structure similar to that of the rural tribes. Often these denizens of our heartless cities occupy the same territories as dispossessed humans and, like them, shelter in cardboard boxes.

Free-living cats from Bangkok to New York form stable social groups whose members reject outsiders and spend copious amounts of time grooming one another, indulging in mutual head and body rubbing, and gamboling and playing together, after which the participants often saunter off for a stroll or take a nap together. Breeding females also make communal nests, sever one another's umbilical cords, help clean and dry the newly born kittens, and bring their offspring up together—one mother always remaining to guard the nest while the others hunt or carry out other important tasks. Kittens from these feline kibbutz mature more quickly than those from one-family nests, making their first lone adventures at least one week earlier. Feral cats live fascinating lives, survive perfectly well without human assistance, and do not require "saving" for their own benefit. To take a self-determining feline used to a complex social life and a large range and incarcerate her, whether in the wooden cat condos of San Francisco or a London cage, is simply cruel. And capturing a feral cat may have repercussions for the environment—for the cats with most suspicious natures will evade capture, which leaves only the truly wild to mate. They will produce an even wilder cat, which will perhaps over generations take on the aspects, not of the domestic *Felis catus*, but of her wild forebear *Felis silvestris*.

City cats share their territory with dispossessed humans and, like them, use newspaper to keep warm.

The Home Range

But what of the adored and free-ranging domestic cat? It would surely be strange if she did not adopt the same *modus operandi* as her feral brothers and sisters. In densely populated cities where many felines live in apartments without individual gardens or any other kind of demarcation zones, domestic cats tend to form colonies very similar in make-up to those of the feral cats.

Pet cats can find it hard to establish home ranges in the heart of the city.

This is particularly so around squares or by leafy churchyards. It is easier for feral cats to hold territory here and the domestic felines, particularly in extremely dense areas with no green space, find forming a home range hard work. But in spacious suburbs she reigns supreme and it is the feral cats who must negotiate for space.

At first sight it seems the pampered suburban cat behaves quite differently from her feral relatives, but in fact her behavior is similar—it's just that in suburbia the female domestic cat considers her humans to be part of her tribe and so the overlapping core area is the home, while the garden becomes her home range. Toms as usual maintain a range that overlaps that of several females. Home range is, however, a movable space. Felines will defend it against some cats and not against others; its size may change according to the weather, the availability of mice, or the discovery of a new delicious place to snooze in the sun; and in suburbia changing alliances or perhaps just the desire to play may mean that sometimes a strange cat is made welcome and becomes an honorary member of the inter-species tribe.

Even though cats will, as the mallard-hunters demonstrate, work as a team if necessary, generally, cats hunt alone—a consortium is hardly needed to bring down a mouse—which means success, or failure, does not affect the well-being of the tribe.

The cat then, is responsible for her own destiny, which may explain why although sociable *Felis catus* is so essentially herself. She doesn't *need* to co-operate with her tribe and, as far as the pet cat is concerned, that includes you, her adoring caretaker. She sits on our laps and luxuriates in our caresses *when it suits her* and claws the self-same hand when it doesn't. Groveling is not in the feline vocabulary, reprimands are met with fearsome hisses, and a crucial indicator of her independence is refusal to keep vets' appointments. No matter how painstakingly feline caretakers disguise the fact that today is vet day, their cats frequently disappear before dawn. Out of 65 veterinary surgeries in north London, 64 agreed that clients frequently canceled appointments because cats had gone absent without leave—in despair, the sixty-fifth surgery had given up an appointment system altogether.

The Story of Reebok

Reebok was a well-known feline-about-town in Los Angeles. Sporting a red heart-shaped ID tag, the charming Abyssinian/tabby cross whose four white paws explain her name, charmed everyone. Always ready to accept an invitation to partake of a saucer of milk or laze on a sofa, Reebok was queen of the walk.

One night when she had "slept over" for 24 hours, the human-in-residence Jane, thought she should call the number on Reebok's tag in case her caretaker was worried. Worried? Peggy was flabbergasted that Reebok was out and about, and that she knew so many people so well. Peggy didn't even live on the same street and more astonishing still was that Reebok was left in a large, bright, garage, with toys, food, even music while she was away on two- and three-day trips. She had no idea that the moment her car pulled away, Reebok opened a screen and was off to be pampered by her many alternative families, because—even more amazingly—Reebok was always back in the garage when she returned.

As time went by Reebok's allegiance to Peggy dwindled and he was mostly to be found with Jane and Oke, where there was always someone home and a minimum of four dinner parties a week, which for Reebok was heaven. She'd invite herself over the moment the party started, stationing herself in front of the low French windows at the front of the house, tail curled around her by now rather rotund bottom, ready to greet the arrivals. On seeing her people would often say, "Is Peggy here?" and Jane would say, "Oh no, she's flying, so Reebok came alone ..." When the last human had arrived, Reebok would leave her post and circulate, never jumping up, but indicating when she was ready to be served a canapé.

Cooperating With Other Species

Just because cats are independent and resourceful, it doesn't mean they are necessarily selfish. One habitually meowed loudly to summon her canine partner in crime when the pantry coast was clear. On one occasion she was discovered "mounted on a shelf, and keeping with one foot the cover of a dish partly open, was throwing down to him (the dog) with the disengaged paw some enjoyable good things."

Inter-species cooperation takes many forms; here one cat serves her canine partner with ill-gotten gains.

And Peter, a tom, if he heard his canine companion whining in his dark cellar prison where he was sometimes confined because of his "dirty habits in the house," always lifted the door latch to set him free. While another, whether displaying empathy or missing a playmate, when hearing his canine companion rap at the door would spring at the door latch, lifting it up, allowing the dog to push the door open.

But do cats love us? Do they feel affection for us in the same way a dog does? Do they crave our company or are we just a feline service industry?

Like us cats have their preferences and can, for inscrutable reasons, leave one home for another, prefer the freedom of the open road, or insistently approach one particular human until they succeed in being adopted. Francis, who already owned a dog and two cats, was amazed when every evening on his return from work, a small, beautiful, and passionate ball of fur, possessed of striking ear tufts, began to hurl herself at his legs and attach herself firmly to his trousers. No other member of his household was solicited in this manner and Friday, who was soon taken in to join the tribe, remained true to her human until death, never sitting on any lap but his.

Other cats appear to risk much to return to a human they loved, as did Sugar, a feline with a unique deformity in her hip bone. When Sugar's humans moved 1,500 miles (2,400km) from California to Oklahoma, they thought Sugar would be happier staying where she was and found her a new home with a neighbor. Two weeks later she was gone. Fourteen months later a disheveled but otherwise healthy Sugar, much to the amazement and happiness of her original caretakers, strolled across their Californian lawn. Skeptics, both of amazing feats of navigation such as this, and the emotions behind them, claim that the humans are mistaken, that the prodigal son is merely a stray bearing a physical resemblance to their long-lost pet but in this case the skeptics were proved wrong when Sugar was unmistakably identified by her hip bone.

Sugar traveled across the US from Oklahoma to California to be with the humans she loved.

Felines surely experience loss and pain, for us and for their homes. This has been vividly described by writer George Moore, who wandering around the smoking ruins of Dublin in the aftermath of the Irish Rebellion, came across a broken wall to which a mantelpiece still clung.

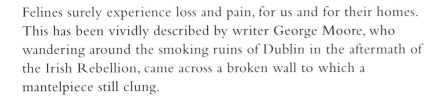

"A plaintiff miaw reached me, and a beautiful black Persian cat appeared by the fireplace. A cat is almost articulate, and Tom asked me to explain the meaning of all this ruin. He has found his old fireplace I said, and tried to entice him; but pleased though he was to see me, he would not be persuaded to leave what remained of the hearth on which he had spent so many pleasant hours, and pondering on his faithfulness and his beauty I continued my search amongst the ruins meeting cats everywhere all seeking their lost homes among the ashes and all unable to comprehend the misfortune that had befallen them."

So what hides within the heart of cat? Love, hate, empathy, coolness, co-operation, selfishness, sagacity, attachment, and jealousy but, above all else, the desire to be true to her essential nature, *simply to be a cat.*

CHAPTER FIVE:

THE PSYCHIC CAT

"Sometimes, alone with a cat in the dead silence of the night, I have watched the cat's eyes suddenly dilate, her ears point back; with arched spine a startling unexpected prance across the floor follows; the puss settles back to laundry and repose as if nothing has happened. What has happened? What has awakened this fit of wildness?"

—Van Vechten, cat devotee (1880–1964)

Mysterious, seraphic, and unknowable, for millennia the cat has been credited with occult powers, imbued with religious and symbolic meaning, and even deified. In ancient Egypt, the dedication "for the beautiful and gracious cat" was a usual offering phrase when addressing the divine Bastet, Sekhemet, Mut, and Neith, while the all-powerful god Re was addressed as "the Great Tomcat." In this chapter we consider more modern, but no less potent, tales of the feline and her powers.

Can the cat see more than we? Certainly intimations of magic pulse from her eyes. Dramatically large, glittering mirrors by night, yet with pupils round and flashing as the sun; clear, wide, and deep by day, yet with pupils slit to a crescent of the moon, they seem to pierce beyond the merely material to realms of mystery denied to mortals such as us.

The cornea and lens, well set back behind the cat's iris, are highly curved, her retina rich with light-sensitive rods compared to color-sensitive cones—man has four rods for every cone, the cat twenty. Behind her retina a crystal layer, the *tapetum lucidum*, reflects back to the rods and cones any light that manages to penetrate the retina, and in dim light crossing fibers pull the iris open to a wide circle. Her field of vision is 205 degrees, ours 180. While we sleep our eyes are covered by thick fleshy lids; when she wishes the cat uses only a third, translucent eyelid—the nictitating membrane—which flicks open at the merest intimation of shadow.

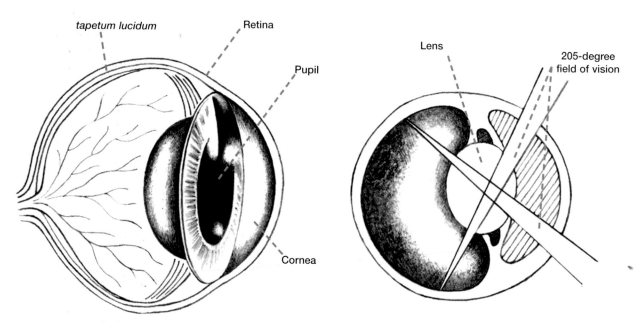

Cat's eye in profile; The structure of the eye allows large amounts of light to be processed (left) while the field of vision is wide (right).

The cat then is visually aware while asleep, and can see in a sixth of the light we humans need and in a 25-degree wider spread.

Does her brain, like ours, make shapes out of the darkness, turning chairs into bears and hatstands into devils, which melt into nothing when attacked? Or does she perhaps see that which we cannot?

A Cat's Revenge

The experimental psychologist David Greene, in his book *Incredible Cats* recounts a story told to him by an old sailor in Malta. This man worked on a Panamanian rust bucket captained by a drunkard who drove his crew hard and who nursed a deep personal hatred for the second mate, a German called Hansen. Shy but efficient, Hansen's only friend was the ship's cat Rhaj, "a scruffy spiteful animal" whom the German fed, allowed to sleep on his bunk, and talked to for hours. In return Rhaj followed Hansen around the ship like a faithful dog.

One night, angrier and drunker than usual, the captain made his way to the bridge to berate his helmsman. Hansen protested, the captain struck him savagely. Reeling, Hansen's head hit the steel bulkhead cracking his skull. Death was instantaneous.

The next evening when the watch changed, the cat came on to the bridge and "stayed staring impassively towards the spot where his friend used to stand on duty." He then continued Hansen's daily routine: "Waking at the same time and proceeding to the saloon before making his rounds of the vessel." The crew became convinced Rhaj was following his friend's ghost. Dread and alarm filled the ship and the captain ordered Rhaj to be caught and thrown over board.

He vanished, only to be found two days later curled up on the captain's face, whom he had suffocated while the captain lay in a drunken stupor. "The strange thing was the lock. The Captain always locked his cabin door from the inside. It was like that when we found it. They had to break open the door to get in. There *was* a spare key—the skipper kept it in a drawer on the bridge. Not many of the crew knew it was there—he liked to have his little secrets, did the old man. But there was one person who, to my certain knowledge, knew exactly where the key was kept. That was the dead second mate, Hansen."

Did Rhaj see the essence of his dead friend, or did fear and hysteria of men trapped with a violent captain lead to fevered imaginings? Did Rhaj inadvertently get locked in with the captain or did Hansen's shade turn the key for him? The cat alone knows the truth and we must make of the evidence what we may.

Cats Foretelling Disaster

Cats have been associated with sailors and the sea for hundreds of years and owe their worldwide spread to being taken along as congenial company on long sea voyages. For example, analysis of worldwide gene-frequency maps of fur color suggests that orange-coated felines originated in Asia Minor and were transported to the north and west coasts of Scotland and the Faroe Islands by the Vikings, probably just because the Vikings liked their looks.

Disaster awaits a ship whose cat has transfered her allegiance to another.

But as a practical seagoing companion, many cats also became full time ships' residents, and not surprisingly a body of superstition surrounds them.

Traditions include the beliefs that a lively cat portends the arrival of strong wind, while the drowning of the ship's cat is an event of unparalleled bad omen presaging the death of all who sailed with her. Crucially for the sailors, cats are also said to know when a ship is making its final ill-fated voyage, and then desert it. A well-documented incidence of feline prophecy concerns the British destroyer H.M.S. *Salmon*, upon which two cats dwelled, the beloved and special pets of the crew. These cats had never shown the slightest inclination to leave their comfortable berth and yet, for no discernible reason, when the *Salmon* docked next to H.M.S. *Sturgeon*, both cats made desperate and wild attempts to board her. They were repeatedly driven off by the *Sturgeon*'s dogs and crew, but would not give up their attempts and managed, as the *Salmon* at last cast off, to land with one final, frantic spring onto the *Sturgeon*'s deck. The cats were proved correct when H.M.S. *Salmon* was disastrously wrecked. Coincidence or feline premonition?

The Ghost Cat

This anecdote concerns cats' most strongly contested power of all, the ability to present themselves to others as ghostly apparitions. This tale was recorded by the vet and naturalist R.H. Smythe in 1959.

"My friend owned a large smooth-coated black tomcat, remarkable for the fact that it possessed a squint in either eye which gave it an extraordinary appearance, as a large part of the white of each eye was revealed. The cat disappeared from home quite suddenly about six months previously and its fate was never discovered. In its place a gray Persian kitten was obtained, which developed into a fine neutered male.

I had once or twice observed Topaz, as the new arrival was named, sit up suddenly and stare at something in the room, then settle down again. Soon other people began to notice this peculiar behavior and talk about it.

A month or so later a traveler, quite strange to the district, called to see my friend about some piece of scientific apparatus he proposed to order and they sat together in my friend's sitting room in conversation for about an hour. During this time Topaz sat up once, ruffled his coat, stared at the mat in front of the fire, then settled down again. Nobody made any comment at the time.

When the traveler was leaving the house, in good spirits after a cup of tea and a good sale, he bent down and stroked Topaz, he mentioned that he liked cats, but of the two he preferred the smooth one.

When asked which smooth cat he was referring he replied: 'The black one that walked in and lay on the mat while we talking. I was fascinated by his large white eyes.'"

Extraordinary, isn't it?

The following pieces of cat magick testify to the cat's ancient abilities to advise in the fine art of weather prediction:

"When Kitty washes behind her ears, we'll soon be tasting heaven's tears."

"A cat who sits with its back to the fire is said to be a portent of frost."

"If your cat sneezes, be prepared for rain: If she sneezes three times, you will catch cold."

The Cats' Special Senses

Phenomena such as cats predicting the weather and even the advent of earthquakes fall into the province of animal folklore; although the cat's seeming psychic abilities may yet prove to have a basis in biology and physiology as we come to understand more about their powerful senses.

The cells of a cat's nose and upper lip convey extremely detailed information on temperature change; she can detect the slightest movement of air as it passes over her fur; it is thought that her face whiskers, or *vibrissae*, are so sensitive that they can even detect the air currents deflected by objects she passes; she can hear sounds too faint for us to catch as well as a large range of ultrasonic sounds the upper limit of which is 60kHZ. Our upper limit is around 20kHZ, top-register rodent squeaks, which are of great importance to *Felis catus*, are around 25kHZ while bat squeaks can reach 130kHZ.

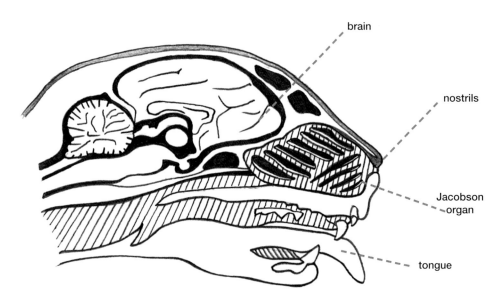

This cross-section of a cat's head shows the size of the Jacobson organ, into which the cat filters air for large amounts of sensory information.

David Green, in his book *Incredible Cats*, tells of Barney, a resident of northern England who habitually slept on top of his humans' warm television until they went to bed. On one occasion he "woke from a deep slumber and jumped to the floor. After standing and glaring at the flickering screen for a few moments, bounded to the door and demanded to be let out. Not long afterwards the tube disintegrated and showered the room with shrapnel-like fragments of shattered glass."

A likely explanation for this behavior is not that Barney saw the future as such, but that he heard unusual sounds above or below the human range being emitted by his "bed" as the blowout became imminent and decided to leave, just as we might get up from a wooden chair that started making ominous groaning sounds.

Felis catus, in common with other felids and some mammals such as the horse, also has a Jacobson [or vomeronasal] organ—which humans do not possess. Despite many studies researchers are still not certain of exactly what range of olfactory and possibly other information it supplies. Positioned just below the cat's nasal cavity and opening to her mouth through a duct just behind her first incisors, the cat has consciously to bring it into play, unlike her eyes or ears, which supply a constant stream of information to her brain. To use the Jacobson organ the cat adopts a grimace-like expression with her mouth partially open, known as the Flehmen response. This closes off her usual breathing route and draws air through the ducts into sacs inside the organ. Having done this she often licks her nose.

Cats certainly use this organ enthusiastically to scrutinize urine marks; males frequently "Flehmen" to assess cats' gland secretions, a female's sexual status, as well as catnip and other heady, non-biological aroma odors. The Flehmen response takes only around a second and is fairly subtle in the domestic cat, making it easy for human companions to miss, though a controled setting can never replicate the smells and sensations of everyday feline life in the streets of New York or the farmlands of Italy. Nor can it introduce all the odors or chemicals a cat might be interested in, so it may be that the cat uses this organ to detect chemical information from sources of which we are not aware and then act upon them. They might, for example, be able to identify the invisible airborne molecular ingredients of an approaching forest fire, before it can be seen, felt, or heard by man, and so be able to relocate.

Can Cats Predict the Weather?

"When cats wipe their jaws with their feet, it is a sign of rain, and especially when they put their paws over their ears in wiping" is ancient weather-lore which prompted Roger Tabor, the biologist, writer, and cat lover, to investigate further. He dismissed the first half of the maxim because cats wash their jaws after every meal, behavior explained thus by a charming folktale: A cat caught a sparrow who astutely playing on the feline reputation for vanity observed, "No gentleman eats before washing his face." To prove his superior status the cat naturally relinquished the bird, who flew to freedom. Since then cats wisely wash after dinner.

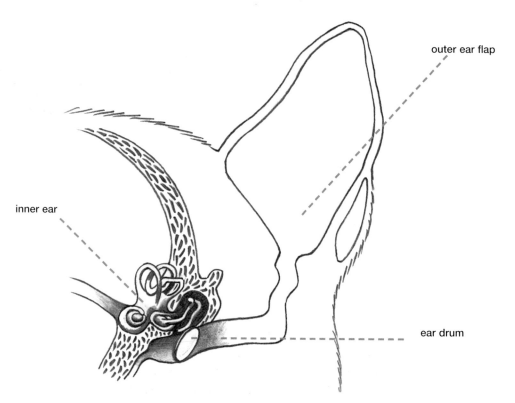

outer ear flap

inner ear

ear drum

This cross-section of a cat's ear shows how a large outer ear flap draws air into the ear, possibly allowing the cat to detect changes in air pressure.

However, commenting on the second line, Tabor wrote, "Every time I have ever known a cat not just to go over its ears in washing but also firmly rub into the little depression in front of each ear, it has resulted in rain by the following day."

He speculated this might be because the cat's ear drum is designed to detect slight air pressure changes—both we and the cat hear through vibration—so humidity changes would also register on the membrane-free surface of the cat's ear, causing a sensation not unlike the human ear "popping" when the pressure inside an airplane changes. Vaguely irritating, this might cause the cat to rub her ears. Fascinated, Tabor conducted a survey of cat companions via the BBC television show *Animal Magic* asking viewers to record "if and when their cats washed their ears and to note how long afterwards it next rained." Results were somewhat patchy—cat companions did not always notice if their cats had washed their ears (cats do it only for around a minute)—or noticed it was raining but had no idea when the shower had started. But, those who did notice both events recorded that ear-washing generally occurred within four hours of rain and more frequently within one hour.

It is recorded that a cat's ear-washing generally occurs within four hours of coming rain, and, more frequently, within one hour.

Perhaps if you keep a log of your cat's ear-washing, being caught in the rain without an umbrella will become a phenomenon firmly relegated to your past. Other cat behaviors said to herald rain include scratching a fence (in a tradition originating from western Maine), washing her face before breakfast (from eastern Kansas), or washing in the parlor (from New England.)

Cats seem particularly to dislike getting wet, even, unless extremely hungry, refusing to hunt in the rain. In summer they use the evaporative effect of their own saliva to cool their bodies so it is to be expected that wet, drenched fur in winter is likely to leave *Felis catus* feeling extremely cold and miserable. Far from conducive to health, well being, and general survival, predicting rain gives the cat an evolutionary advantage that allows her to take shelter in a warm nook before it arrives.

Cats, along with other animals, have been said to know when earthquakes are imminent.

The cat is also credited with being able to bring rain, particularly in agricultural economies and subsistence farming communities where the vagaries of the weather can mean the difference between life and death. In Malaya, women first place an inverted earthenware bowl on their heads, then fill it full of water, and wash a cat in it until "she nearly drowns" to ensure heavy rain; while in a village in Sumatra women wade into the local river and douse one another with water. A black cat is then thrown in to join them. Finally, the cat is allowed to escape all the while pursued by splashing women.

Can Cats Predict Earthquakes?

Since the earliest of times, cats, along with other creatures, have been said to know when earthquakes are imminent. The historian Diodorus Siculus, writing about a catastrophic earthquake in which the Greek port of Helice disappeared into the sea in 373BC noted that five days before the quake, much to the mystification of the city's human inhabitants, animal denizens, such as weasels and rats, fled.

Mainstream western scientists seem uninterested in investigating this phenomenon. However, if people who lived in earthquake areas formed a network to report unusual behavior in both their pets and wild creatures, quake prediction could become a possibility, something which seismologists are as yet unable to do except in the very short term.

However, Chinese scientists have been prepared to take a broader view and have since the 1970's used observations by members of the public as an animal early-warning system. By combining this information with other seismological precursors of quakes they have had notable successes in prediction, such as the evacuation of Haicheng city before the quake in February 1975.

Rupert Sheldrake, a biologist interested in animal earthquake prediction, studied animals' reactions over a period that took in the major tremor centered on Northridge, California in 1994, and the Greek and Turkish quakes of 1999. He found that there were many reports of peculiar animal behavior—including "dogs howling in the night, caged birds becoming restless, and nervous cats hiding."

Geologists are quick to dismiss these reports, alleging that people remember only odd animal behavior after a quake and would have thought nothing of it otherwise. Sheldrake disagrees. "Comparable patterns of animal behavior prior to earthquakes have been reported independently by people all over the world," he said.

"I cannot believe that they could all have made up such similar stories or that they all suffered from tricks of memory."

Many quakes are either preceded or followed by strange geophysical changes, like the *tsunami* which devasted south-east Asia in late 2004. In 1975 in China's Haicheng there were areas of high temperature on the earth and "part of the ice in the shade of a frozen reservoir melted during a very cold winter," which must have been the reason that many snakes came out of hibernation and, as a consequence, froze to death. Sheldrake suggests the most plausible explanation for these strange events is that animals respond to the changes in electrostatic fields that occur before many quakes and "probably arise from changes in seismic stress in rocks." If you live in a quake-prone area, why not keep a log of unusual behavior in your cat? Most cats seem to demonstrate fore-knowledge of quakes either by running away or hiding, so look for these behaviors particularly.

The cat, as all her companions know, is a creature of great mystery, power, and intelligence. Closely observing her behavior and correlating it to other events may prove your very own feline to be more magical than you had ever imagined.

The Mysterious Powers of Cats

Cats are ascribed with mysterious powers throughout the world. In Thailand the ancient cat treatises (see Chapter 1) reveal that:

"Any cat with four white feet—two ears
yellow eyes like gold, genuine
who feeds will amass—much treasure
Any action, yet small, brings results."

And the treatises also point out that merely owning any of the approved 17 cat breeds (see Chapter 2) means that there is no need to be "afraid and troubled." However: "Once you have the magic cat, you must arrange for it to live well. Hurry to bathe it, rubbing powder and sandal wood. Do not kick it or strike it. Hurry to put plates of gold and silver for it to eat as appropriate."

The Japanese maintain that ten-year old cats spread a mysterious light into the night. But this particular belief surely has its roots in reality: Static electricity accumulates in feline fur, producing dramatic, sparkling, intense balls of vivid-colored electricity. Easily produced by deliberate friction, or even speedy movement, through thick, tangled undergrowth, the phenomenon is intensified in frosty weather, causing Gilbert White, the naturalist, to write of a cold spell in 1785: "During those two Siberian days my parlor cat was so electric, that had a person stroked her, and been properly insulated the shock might have been given to a whole circle of people."

Cats and Witches

As even stroking a cat gently can result in a thrill of electricity running up our arms, it is no wonder witches pretended they could turn themselves into these magical creatures. At no time was this shape-shifting more vital than at Halloween. In Celtic, pre-Christian times, when the day was known as Samhain Eve, this enthusiastically celebrated harvest festival also marked the mysterious passage of one year to the next, an enchanted time when barriers to other realms dissolved, allowing ghosts and spirits to walk the earth.

Witches attended festivals such as Halloween either accompanied by their black cats, or in the form of a black cat.

The Church attempted to eclipse Samhain Eve by introducing All Hallows Day (a celebration of Christian martyrs and saints) on 1 November. When this strategy failed, they turned to demonization and eventually succeeded in turning magical Samhain Eve into an evil Sabbath where Satan was worshiped in the form of a gigantic cat with lustrous black fur.

Witches attended these Sabbaths either accompanied by their black cats or in the form of one. To effect this magical shape-changing (at least in the witches' minds), ointments containing ingredients

A Spell for Turning into a Cat

Gathering together arcane ingredients was clearly too much for many witches and they opted for a simple spell, which those who wish to get inside the soul of their own cat might like to try out for themselves at midnight on Halloween:

"*I shall goe intill ane cat*
Wi th sorrow, and sych, and a blak shott."
To return to human form merely recite:
"*Cat, cat God send thee a blak shott*
I am a cat's likeness just now
But I sal be in a woman's likeness ewin now."

such as ground toad, serpent, and fox, and the reputedly hallucinatory drugs belladonna and mandragora, were mixed with blood and spread two inches thick over their bodies.

But why was it that the Church chose cats to metamorphose into evil beings? Because cats were associated with the extremely popular pagan Moon goddess cults of Diana, Artemis, and Isis. The ancients also held the cat sacred to the Moon. She embodied mystery and, as her eyes mirrored that heavenly body's properties, she quite naturally became the Moon goddesses' sacred animal of divine incarnation.

Because Diana's following was particularly strong in Britain, the Church found it imperative to debase her. The Moon goddesses became powerful witches and their mystic cats satanic familiars who would perform "small malicious errands, including murder." These were dangerous times for pet cats, particularly those who were black and owned by ladies of certain years.

Supposed satanic and pagan links made the
Medieval age a dangerous time for black cats.

Music and Emotion

But what of the flesh-and-blood pet cat? Is she really imbued with extrasensory perception? Can she communicate with us in nontactile and nonverbal ways? And can we reach her without speech?

Mood and character are created by complex mixtures of emotion, physiology, genetics, psychology, and external forces. Music also elicits powerful emotions by activating the cortical systems associated with them and also produces a host of subconscious activity in the parts of our nervous system that in evolutionary terms are most primitive.

At first glance emotion induced by music seems to have no particular benefit for our survival, but it can help provide social cohesiveness—a contemporary example being the singing of British soccer tribes. Music can heighten arousal and induce bonding—many peoples have used music to inspire their warriors before battle. Conversely it can lower arousal by diminishing testosterone levels, so reducing group conflict and sexual competitiveness. A baby's survival capacity is dependent upon the strength of its bond with its mother; a mother singing or "talk singing" to her infant before it can express itself verbally intensifies their closeness. If language is part of the mind's general facility for learning and memorizing and as such, not the exclusive province of humans, then, if music is language's precursor, it is reasonable to assume that animals also have the capacity to react emotionally to it.

The cat has been associated with music since the days of her Egyptian pre-eminence when her lithe form decorated the sistra— a brass-framed instrument of loose metal bars, which produced a sound like a tambourine when shaken. Moncrif, a French feline devotee writing semi-seriously in the eighteenth century, made a case for cats singing at Egyptian banquets. Although the divisions between feline notes are small, their language is tonal and is

The cat has been associated with music since ancient Egyptian times.

musical. Champfleury, author of the seminal tome *Les Chats*, detected 63 different mewing notes. None of us can deny the harmony of a cat's trill and some composers, such as Carlo Farina, seventeenth-century court violinist at Dresden, even made the cat's language into music for us.

Do Cats Enjoy Music?

They certainly delight in walking over keyboards, reveling in the sounds they can produce, yet mystified from whence they come. Feathers, a Persian, "walks sedately from one end of the keyboard to the other, producing an exotic succession of sounds." And tradition has it that Scarlatti's *The Cat's Fugue* was based on his own feline's keyboard meanderings.

There are also numerous accounts of cats seemingly choosing to listen to music. One of the most curious examples is that of celebrated harpist Mademoiselle Dupuy's feline, who sat next to her while she played, registering pleasure or annoyance depending on the skill and charm of her human companion's performance. Mlle. Dupuy felt she owed everything to her cat's superb musical judgment and, on her death, left the cat her estate.

Tradition has it that Scarlatti's *The Cat's Fugue* was based on his own cat's keyboard meanderings.

Whatever the truth of this, contemporary research conducted by Professor Bubna-Littitz at the Vienna Veterinary University shows clearly that cats have an inbuilt sense of rhythm and like an insistent beat accompanied by deep tones.

The compilers of CDs that are specifically designed to improve behavior, and reduce stress and separation anxiety in pets, use the theory of entrainment. This postulates that two rhythms side by side—here, the cat and the musical vibrations—will gradually begin to move at the same rate. By selecting classical music to be in line with feline biorhythms, and adding ocean wave sounds at a frequency emulating feline Theta waves (which affect the cat subliminally), the cat's disposition is greatly enhanced. The music itself puts her in a good mood, while the biorhythms relax her—an optimum state of mind for learning and feeling good.

The Power of Telepathy

The existence of paranormal powers and telepathy are fiercely contested by the scientific community. Skeptics ask what evolutionary purpose they serve.

In an urban society the justification of the existence of telepathy is not obvious, but undoubtedly an ability to tune in and resonate mentally with other creatures is a mental skill that enhanced our survival as hunters and gatherers, and could similarly bestow great advantages on other animals. Prehistoric man understood in an absolutely direct way that he could not survive alone, that without the compassion of the essences of earth, sky, and water, which surrounded him, his life force would be quenched, and without the animals that also surrounded him he would die.

Are Cats Telepathic?

R.H. Smythe, a London naturalist and vet, possessed a beautiful Siamese, Wun Lung, who in essential cat style, treated him with the greatest of indifference whether he was telling her off or offering her a plate of salmon. But this exterior nonchalance masked a tremendous passion as she "refused to leave my side except under the influence of great force and would then escape and track me down. Even when miles from home, attired in city dress and laden with umbrella, or engaged in someone's office or inner sanctum on the most confidential or urgent business, a loud and penetrating cry would announce the arrival of Wun Lung. The whole thing became such a worry that my friends remarked on my constant appearance of apprehension and the habit I had developed of casting furtive glances around me."

How could Wun Lung have known where her beloved human was except by telepathy? It seems even more far-fetched to imagine she could have tracked him through a busy city by smell, when a cat's olfactory powers do not begin to reach the acuity of the dog.

Try to discover if your cat is using her sixth sense to greet you on your return home.

From this need came the shaman, whose spiritual authority derived ultimately from his or her ability to communicate with guides and guardians who are the spirits of flowers, animals, and elements or gods, and whose prime social responsibility was to secure the success of the hunt. Without this, his people, in common with other predators, including the rural feral cat, would starve. The shamanistic world view sees no divide between the soul and being of human and beast—they are kin—and from this point of view it seems perfectly natural that our cats might understand the essence of the mice they hunt and feel our thoughts, whether we be near or far.

Contemporary technological society, which is based on speed of communication, has, in evolutionary terms, hardly begun. We possess within us the same abilities as our tribal ancestors—the shaman was a specialist in communication, but all tribal members exercised these abilities. Perhaps we just need to learn to use them again? Powers that were once taken for granted are now regarded as fantasy, because on one level contemporary society no longer needs them, and on another because people don't wish to be associated with what is generally considered to be primitive or "unscientific."

There are many reports of cats knowing when their caretakers are returning. Rupert Sheldrake, the biologist who has researched this phenomenon extensively, has collected 359 accounts of this behavior. In a random survey of 1,200 households in the UK and the US, he found 91 whose felines seemed to know when their owners were coming home. His research on this phenomenon in dogs is well documented and, of course, it is relatively easy to video canines while home alone as they generally wait in one place, usually by the door or window. The freeranging feline, however, may choose to wait in a number of places, depending on the weather and her mood, making filming an impossibility. Dr. Sheldrake suggests that instead, interested humans make a log of their cat's activities. This is particularly valuable if one person in a couple documents the cats behavior in relation to their partner's comings and goings. Csaba Pasztor, companion of luscious thick-furred Cica reports that whenever he returns, be it by foot or car, Cica is there at the gate to greet him and often it is clear she has just woken up, meaning she must have awoken just for him.

I have fed a neighbor's large and somewhat voluptuous marmalade cat on dozens of occasions when his humans are away. If I am prevented from feeding him at his usual time, at least I know Tom will not be desperately hungry, as there is usually some food left in his bowl. Yet, on every occasion except two, when I finally arrive he is waiting at his front door, mewing pitifully for me. Of course I have no way of knowing how long Tom has been waiting for me, but I watch his empire from my window and know he spends much time patroling the large gardens which make up his home range, sleeping on the shed roof and relaxing on his balcony, so he is clearly not waiting for me constantly.

There is much anecdotal evidence of cats being aware of significant events at a distance. Sheldrake recorded the case of Jean Parker, whose cat Timmy was generally to be found snoozing on her son's bed when she returned from work, but on one fateful evening was instead meowing pitifully. Initially, Jean thought "he was in pain, but no amount of fussing would calm him down." But when at 8.15pm she received the news that her son had been in an accident and was in a coma, the reason for Timmy's distress became evident. Jean's son stayed in a coma for seven long weeks and not once did Timmy go into his bedroom. Then, "one evening Timmy ran straight into my son's room jumped on the bed and began purring with deep pleasure. That was the day my son came out of his coma and began to get back to life again."

Cats sensing major events that happen to humans is relatively common but, as Sheldrake discovered while gathering case studies, the reverse is extremely rare. And, curiously, Dr. Osis who researches cat–human telepathy at Duke University and found cats quite receptive, commented: "My biggest regret is that no cat in the experiment showed the slightest interest in sending a telepathic message back to the scientists."

A loving cat rolled on his master's bed as the boy came out of a coma.

Is Your Cat Telepathic?

Calling silently is a way of discovering if your cat is open to telepathy. If your cat is freeranging, call for her at random moments and see if she comes to your command. Keep a log of when you call her and how often she appears and the period between your call and her arrival. If she appears soon after your call in a number of tests, but does so significantly less at other times, she may well be responding to you—although cats, unlike dogs, are quite capable of ignoring your call, which makes evaluating this kind of telepathy in felines somewhat tricky.

It may be the cats have little to say to us—even in the material world their communications with humans are principally concerned with obtaining dinner or being the other side of a door—but perhaps it's all a matter of being open to the feline psyche. If you think your cat has sent you a message, take a chance, respond to it, see what happens. It may open an extraordinary interspecies communication, far more meaningful than mere speech. One day, it may also save her life.

In contrast, US psychic Jeanne Dixon has no doubt that it is quite possible to have perfectly lucid two-way telepathic conversations with cats. While in a meeting with her, some Japanese businessmen were on the verge of leaving the room—a terrible loss of face for the host who had invited her over. At this moment a woman, accompanied by a Turkish Van, entered. The cat sat itself on Dixon's lap and she immediately began to converse with it but "had forgotten that the cat … was accustomed to hearing Japanese. As I waited expectantly for answers, the cat looked at me with similar questions written all over its face. 'Who are you? Why do you speak so strangely?' In this case, the foreign language and my unfamiliar scent confused the cat at first, so I spoke again: 'Tell me what is it? What are you?' The cat answered by licking my hand. 'Thank you,' I said out loud while sensing the full attention of the disturbed committee. To the cat I whispered: 'Can you get me out of here?' With that the cat began to emit a series of sounds that included chirrups and gentle mews. Her message was clear. The cat told me that the men in the room were as fearful of the outcome as I. I looked up at them and knew that the cat was right!"

Paranormal skeptics are convinced that pet owners suffer from advanced cases of wishful thinking. James Randi, a conjuror, is so certain his beliefs in the psychic power of animals are correct that he has offered one million dollars to anyone who can demonstrate a psychic phenomenon to his satisfaction. If you believe that your cat is telepathic and will perform to order in a laboratory setting, why not challenge him?

The Power of Staring

Although it remains unproven, the act of staring is another mode of communication; a high percentage of adults and children believe in the sense of being stared at, a belief that increases with age. It may be that as we get older, the cumulative number of times we actually experience this phenomenon overcomes our ingrained, scientifically-based skepticism.

Discover if your feline has the power to know when she is being stared at.

Scientists make much of the fact that rays are not emitted from our eyes and it is thus impossible that we can sense when either another human or another creature is watching us. Science took a very long time to discover that bats navigate using sound waves (echolocation), or that butterflies can see ultraviolet light rays, but these abilities have always existed, their mechanism plain only once discovered.

It has been postulated that through focusing our visual attention on something at a distance it is as if our mind extends outward to connect us with this distant object. Although our mind is focused like this, we are still aware, to a greater or lesser extent, of the environment around us, and this general sense of environment is projected into a perceptual field which is both in and around our bodies. When the perceptual fields of starer and staree meet, the staree's field alters and he experiences the sense of being stared at. In an environment potentially charged with danger, be it a dark alley or a jungle, we—and cats—are alert to every nuance of change around us and would immediately notice this difference.

In an informal survey Rupert Sheldrake asked people if they had ever found that they could stare at an animal from behind and make it turn round. Fifty-five percent answered yes. Why not experiment on your feline? You may be surprised at the results.

The Cat of Your Dreams

The meaning of dreams has exercised the minds of humankind since the beginnings of time. Most cultures have attempted to interpret the many creatures that populate them, whether human, animal, or monster, and divine meaning with reference to their own folkloric or mythical world, as well as the dreamer's own reality. This tradition has led to rather barren, stereotypical interpretations of real animals. For example, the black cat's reputation as a witch's companion has ensured that she is regarded as a figure of ill-omen, while universal feline identification with the illustrious, benign Egyptian goddess Bastet means she can also represent the apotheosis of feminine power and the divinity within.

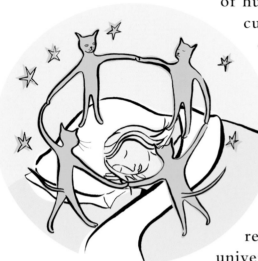

As in life cats in dreams can come to help and heal you.

Other dream interpretations based on society's ambiguous perceptions of the flesh-and-blood animal mean her langorous, uninhibited sexuality can be variously seen as a symbol of fecundity, depravity, devouring sexuality, or vice, while her independence of character and paradoxical behavior is critically categorized as "catty."

Psychoanalysis, as practiced by Freudians, neo-Freudians, and Jungians, has devoted itself to symbolic interpretation, but even today has reached no consensus on what dreams mean. Freud considered the cat to be a symbol of the genitalia, but, to quote from G. William Domhoff, "In the case of Freud, not a single hypothesis put forth by him on any topic related to dreams receives empirical support in the dream literature." Therapists tend not to discover the actual meaning of their clients' dreams but "through the process of suggestion and persuasion" establish "what may be a common belief system." So where does that leave the cat of your dreams? Domhoff and his colleague, Schneider, at the Psychology department of the University of California, Santa Cruz, have analyzed thousands of dreams by content and pattern, correlating them to the lifestyles and interests of the dreamers. Their website (www.dreamresearch.net) contains a database of over 10,000 dreams, which can be analyzed in many different ways. It reveals, for example, that animals occur in the dreams of young children 30 to 40 per cent of the time, in American men 6 per cent, and women 4 per cent. Animals appear in as many as 30 per cent of dreams in some hunter-gatherer societies. Domhoff undertook further analysis to establish whether "people's concern with cats and dogs could be predicted on the basis of their dreams." The dreams of Alta and Barb, two cat-lovers, were populated by felines in 13 per cent and 5.6 per cent of cases. So perhaps dreaming of your own cat, or even other cats, is just a reflection of your day-to-day life's passions, concerns, and worries of which a beloved pet is an important part.

Dreams that Heal

A dream is a jigsaw puzzle and needs to have all its pieces in place before its full significance can be divined, as this example shows. Anita saw herself as an unlucky person in life and was worried that the presence of a black cat in her dream foretold further problems for her. For the full dream, and other cat dreams, consult www.dreammoodscom.

In Anita's dream her daughter appeared with a black cat. Anita asked her if she had fed the cat, and they took it to the bathroom where they tried to clean spaghetti off the sink. Her daughter then put the cat into a bowl full of milk. "He did not drown, he was just standing up in the milk and we were looking at it."

Jade, the dream interpreter, pointed out that, because Anita so firmly believed she was unlucky, she herself was embodied in the black cat—generally in the West a symbol of bad luck—so the dream feline had no additional unfortunate meanings. Jade continued, "It is no coincidence that the cat is soaking in a tub of milk. Milk is symbolic of maternal instincts, nurturance, and motherly love. And hence your dream may be a reassurance that you are surrounded by love."

111

CHAPTER SIX:

TOUCH AND HEALING

"I love cats because I enjoy my home; and little by little they become its visible soul."

—Jean Cocteau, 1889–1963

One of the most powerful senses, touch can delight us, help us, and even heal us. Physical and emotional security are taught to us as newborns by the power of touch, whether it be caresses from a human mother or the furry embrace of a fraternal kitten. As they grow up cats have extraordinarily acute powers of touch, with paws that may even be able to feel changes in the earth's movements. It may be this sensitivity that helps cats become our healers too; the power of pet healing is now proven, with benefits from massage and pet contact improving the health of humans and cats alike.

Our first feelings of security come from the physical closeness we have with our mothers and so do the cat's. Feral nursing mothers share nests, so from the first moments of their lives, when (because they are blind) touch is everything, they know the comforting feel of their siblings, "aunts," and myriad other kittens. Snug in their nest, for the first three and a half weeks of their lives, their warm furry bodies jumbled together, they have everything they need; warmth, affection, and milk, whose production they stimulate—just like tiger cubs—by pressing around the nipples with their strong, already agile, paws. At the same time they establish a "teat order," which may indicate which kittens will be top felines later in life.

A tigress will teach her cubs to stalk peacock and hone their hunting skills on birds, mice, rats, and rabbits.

Soon kittens venture from the nest to play, to gambol, and to groom one another, and, most importantly, to hone the skills of the chase, accompanying their mothers on hunting forays for birds and mice, rats, and rabbits. As the tiger matriarch teaches her cubs to stalk peacocks and buffalo, and shows them the secret hiding places of their prey and where the cool water holes are, so does *Felis catus* show her kittens the less exotic denizens of her natural environment. Tiger mothers, with safety in mind, disable large prey such as deer by hamstringing or biting into the muscles of the rump, before presenting them to her cubs to practice the vital kill, as does *Felis catus* with fearsome large rats.

Researchers believe that between the second and seventh weeks of life, this rich assortment of natural *stimuli* accelerates the spread of the kitten's neural connections, which, among other developmental processes, access the area of her brain that controls her social behavior—or, if you like, dictates the kind of creatures she will choose as her friends. Feral kittens are unlikely ever to be touched and cuddled by humans during this period, which is why, although they will approach us, live in our territory, and even allow us to feed them delicacies by hand, they avoid our touch rigorously, although they find consolation, comfort, and friendship by curling up with and even just lying next to one another. Kittens that we pick up and cuddle daily during this period—the optimum amount of time appears to be about an hour a day—place us firmly in their social circle, allowing us to stroke and caress them, while they rub against us to create a tribal scent, as they would with feline group members.

There is also speculation in scientific circles that even the differences in the way cats like to be stroked—some, for example, favor curling up on a welcoming human lap to receive our fondles, while others adamantly refuse this but adore being stroked as they rest next to us—may also be a result of exactly how they were handled as kittens although there are clearly other factors such as temperature to take into account. Some longhaired breeds may simply find our laps too hot too handle, particularly in winter, while shorthairs might find that simply blissful.

Young kittens that we cuddle daily will place us in their social circle, allowing us to continue to caress them in adulthood.

Joining the Cat Tribe

We might imagine that simply being surrounded by humans during kittenhood would also incline *Felis catus* to place us in their tribe, but research shows quite clearly that it is only the magic of touch that can grant us this honorary membership.

When cats sinuously rub against our legs, they first deposit a fatty, highly perfumed substance from their chin and lips (known as the perioral area), followed by scents from their cheeks and temples (the temporal area) and from the base of their tail (the caudal area). But in day-to-day life with other cats, felines use their chin and lips to mark objects and their temporal area to mark people, reserving the potent perfumes of their caudal area to indicate a sensuous sexuality.

Researchers Susan Soennichsen and Arnold Chamove became curious at these specializations and wondered if *Felis catus's* constant proximity to us had altered the temporal gland's use until its sole function was to rub against humans "as a demonstration of attachment and a way of achieving security." If this were the case, would felines find it particularly pleasurable to be stroked there? They cajoled nine humans to stroke their felines three times at each of the three scent gland sites and one non-scent gland site over a total of 12 sessions and to note the results. Stroking the temporal region resulted in three times more pleasurable feedback from the cats than all the other areas combined, while the caudal area garnered the most negative feedback. Some of the humans also commented that as the experiment progressed they particularly enjoyed stroking their cat's temporal region. It is interesting to speculate why this might be. Perhaps as the potent scents penetrate our skin and their biochemical messages reach our brain, they elicit a feeling of pleasure. Or perhaps it is just that the positive feedback itself makes for a more gratifying experience. Whatever the underlying reasons, temple stroking seems to induce bonding, so next time your cat comes looking for affection, why not conduct your own experiment?

How Cats Experience Sensation

The hairless skin of the cat's paw pads is extraordinarily sensitive. Sensors in the skin that covers a deep fatty layer respond to firm pressure, sending feedback to the brain and allowing her to adjust her balance and posture, while other cells register texture, size, shape, and temperature. The cat uses her paws to examine objects closely, and some researchers believe she can also pick up minute vibrations, endowing her with another mode of "hearing."

Combined with picking up other phenomena, such as changes in electrostatic fields, this may allow her to anticipate earthquakes—and certainly there are many reports of cats behaving strangely before a quake hits (see Chapter 5). Preliminary findings by Dr. Rupert Sheldrake and his colleague David Jay Brown, who are investigating this phenomenon in earthquake-prone California, report that cats become nervous and disturbed before quakes and often run outside or hide.

This exquisite sensitivity probably explains why few felines enjoy having their paws touched and forcing the issue may result in a substantial swat—claws extended.

The cat also experiences sensation through her fur—her hairs being embedded in follicles richly endowed with touch receptors, which register the slightest whisper of movement — and her skin, which is covered with tiny bumps, between 45–160 per square inch (7–25 per sq. cm), stimulated by the merest of pressures. This double sensation may be the reason some, even perfectly socialized, cats, hate being massaged or even stroked—although, given time and patience, some, but not all, come to enjoy the sensation and actively seek it out.

How to Massage Your Cat

Massaging a cat is not so different from massaging a human friend. Although she looks very different from us, her basic structure and anatomy is fundamentally the same, meaning she appreciates many of the sensations we do.

Some cats, also like humans, are not always in the mood for massage; they may have important matters of their own to attend to elsewhere or simply dislike feelings of vulnerability.

It is vital never to force your feline to submit to massage. Chasing her round the house, cornering her in the bedroom and holding her down on the sofa, will put both of you in a bad mood and next time she sees you coming, all you will see of her is the tip of her tail disappearing through the cat flap.

Instead, choose a time when your cat is resting contentedly on your lap, in that comfortable niche in the laundry cupboard, or on the sofa where sun streams gently through the window, and is drowsy and amenable. Put on some soothing music, talk gently to her using long, continuous, flat, or descending sounds, and start with the least invasive of strokes, passive touch. This is the technique in which the hand, palm down, is put on the muscle for between 30 and 90 seconds, warming the body and calming the soul of even a jumpy cat. This simple action can be particularly useful in allaying the fears of an abused, nervous cat but may need to be employed for several weeks, even months, before she will accept any other techniques.

If the cat shows any signs of discontent, which include, but are not limited to, ears flattening back, tail thumping, scratching, or her back sinking low to the ground, or tries to move away, stop at once. It is important your feline remembers this experience as pleasant, not overpowering; there will be plenty of other opportunities to try again.

When your feline is happy with passive touch you can begin to approach her more positively. First, gently offer her your hand to sniff, touch, and accept. If she responds by rubbing against you, touch her gently on the shoulder then pick her up and move her to a favorite place, again respecting any signs of irritation by immediately stopping and allowing her to leave.

A basic stroke of massage is known as effleurage. This is a series of long, flowing, gentle strokes, not so different from the movements you make when you stroke your feline companion ordinarily. Starting just behind her ears, using your whole hand, move down the entire length of your cat's body toward her tail, making sure the pressure is even on either side of her spine. Slow and light effleurage is soothing; while deeper, slower effleurage will help drain her lymph and blood systems. Stress and tension will melt away, the mouse that got

Use petrissage gently and evenly on the sides of your cat's spine.

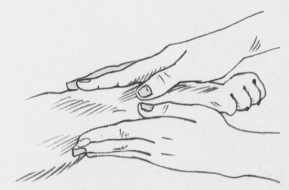

Light effleurage strokes along the forearm and paw.

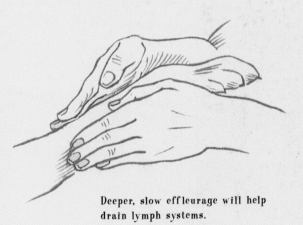

Deeper, slow effleurage will help drain lymph systems.

away will be forgotten, and she may even fall asleep. Faster strokes, however, stimulate muscles and can be irritating, so you may do best to avoid them.

Petrissage is a second basic technique in which pressure and relaxation alternate in a smooth rhythm, which, when slow is soothing; when faster, stimulating. Kneading and squeezing muscle gently presses tissue into the body, while rolling skin and pulling the muscle moves it easily away. This is a great technique if your cat has been very energetic and her muscles seem stiff or knotted.

The Tellington TTouch

Tellington TTouch (abbreviated to TTouch®), developed by championship horsewoman Linda Tellington-Jones, is different from conventional massage. Instead of working on the body's muscular system, it concentrates on moving the skin gently with circular movements and lifts of the hand. These motions activate the healing potential of the body by working at the cellular level, thus both releasing fear and enhancing function.

When living creatures—be they feline or human—are tense, frightened, or in pain, their bodies develop specific habitual responses. Neuroscientists such as American Dr. Candace Pert are beginning to uncover evidence that physical and emotional traumas are trapped in our body cells and conveyed to our brains by neurotransmitters.

Tellington-Jones first worked with the creatures that surrounded her—horses—many of whom were labelled "aggressive" or "resistant." She discovered that when she relieved the underlying cause of the horses' antisocial behavior (usually pain or discomfort) by touching and moving their skin in certain ways, their personalities were transformed. Since its formal beginnings in the 1970's, TTouch has been used to treat creatures as diverse as big cats, non-human primates, and humans. Indeed, taking a TTouch session yourself will give you a genuine insight into how it will help your cat.

How to Practice TTouch

The fundamental movement of TTouch is gentle and flowing. The practitioner places their hand lightly on the cat, gently bending the fingers so that the pads rest on the cat's skin. Circles are usually made in a clockwise direction with a light touch and relaxed wrist pushing the skin in one and a quarter rotations, beginning at six and finishing at nine on an imaginary clock face. Generally, the lines of circles should follow the lines of the body, and the fingers should slide along the skin, connecting each circle and never losing contact with the skin.

If a cat is exceptionally nervous, frightened, or in great pain, try making the circular TTouches with the back of the hand to gain the cat's trust more readily. Cats will usually have one or more body areas that they are reluctant for you to touch. With the circular TTouches you can releases fear or pain in such areas and change behavior. As trust builds between you and your cat, she will gradually let you touch those areas, allowing you to release the hurt.

TTouch can break many common problem behavioral patterns, such as motion sickness, but more importantly, it can also be utilized by vets and behaviorists who may be disposed to use drugs to treat even quite minor psychological and behavioral problems in cats. Medication can prevent cats from exhibiting a fear of fireworks or the nervousness resulting from abuse in early life, but it does not tackle underlying causes. Because TTouch releases locked-in fear and tension, it alters the fundamental way your cat feels about great bangs and vivid displays of color, or her past frightening experiences. In short, she may no longer feel frightened or anxious.

Tellington Touch consists of circles of movement, made with the fingertips. Each movement is one a quarter circles, in a clockwise direction. The movement begins at "six o'clock" (as if on a clock face) and finishes at "nine o'clock."

Aromatherapy for Cats

Massage is very beneficial for your feline; it tones her muscles, alleviates stiffness, improves her circulation, cleanses her system, and greatly strengthens your mutual bond. As you get to know your cat's body, you will notice any small change that will allow you to seek early veterinary advice.

Gentle effleurage (see pages 118–119) helps old, relatively inactive cats to keep their muscles flexible and imbues them with a sense of well-being. In combination with aromatherapy, it can alleviate the pain of arthritis and rid the cat's systems of the toxins which so aggravate the condition. Mix 4 drops of rosemary oil, 2 drops of lavender, and 3 drops of ginger in 1 fl. oz. (30 ml) of a light vegetable oil such as almond. Work the oil through the coat and into the skin with a rhythmic movement, starting from the haunches and moving inward, covering the legs and all the vertebrae. Your cat will lick off any excess oils, ensuring that they also reach her digestive system. For help with rheumatism, try this recipe: Mix 5 drops of ginger oil, 2 drops of rosemary, and 5 drops of chamomile in 1 fl. oz. (30 ml) of vegetable oil.

When Not to Massage

Although massage is wonderfully helpful in many situations, never use it if your cat has a fever; a skin fungus; any infectious disease; neuralgia; acute sprains, torn muscles, bruises, or inflammatory conditions; or is in shock. Also, be careful if you discover a "hot spot" or sensitive area when massaging your cat—work extremely carefully around it. If your cat seems in any distress, consult your vet at once. If your cat has cancer, or any type of tumor or other medical condition, you should also always check with your vet first before starting a massage. Never massage a wounded area. If your cat is too sick or elderly for a full massage, try simply holding their paw flat in your palm; this seems to be much appreciated.

Why We Need the Power of Touch

Touch is powerful yet it is all too often excluded from our lives. Humans who live alone and do not have a partner, may go for weeks, even months, without touching another being, and even those of us who live with a significant other can feel nervous about asking for a hug. Will our partner think we are being "silly" or "childish" and refuse? In our hearts we know how important touch is, yet it can take the most dire tragedy before we will spontaneously reach for a friend and embrace them to give them comfort and strength.

It is surely no surprise that so much contemporary research shows that stroking cats, and other friendly creatures, reduces blood pressure, and that cohabiting with a pet radically increases the life

Holding paws; the power of touch is important to humans and cats alike.

Lynch's Experiment

Fascinated by the correlation between touch and health, Lynch decided to experiment with his daughter Kathleen and her beloved Schnauzer terrier Rags. Kathleen sat quietly for three minutes and her blood pressure remained at its base level; then she read poetry aloud for two minutes and it increased greatly; she was quiet again for another three minutes and it returned to base. Rags was then put on her lap, and as she began to stroke him, her blood pressure fell by almost 50 per cent from its peak to an entirely new and much lower baseline level. In essence stroking Rags created a Physiology of Inclusion, a "biological state of enhanced relaxation which draws people out of themselves and closer to others and the natural living world and which has a profound effect on people's hearts and blood vessels." Your feline can certainly work the same magic.

span of coronary patients—four times as many petless patients died in the year after leaving hospital than those who cohabited, according to research by Erika Friedman, from the University of Maryland. What people are less aware of is the reason for this; it has been named the Physiology of Inclusion by Dr. Lynch, one of the ground-breaking researchers in this field.

What Lynch and his colleagues discovered was, in essence, that many people, even if they are not conscious of it, do not feel they belong to the society around them or the greater living world, and so react to everyday situations, such as talking to a neighbor or hearing a dog bark, as if they were a threat. Consequently their body gears up for flight or fight, which means blood pressure and heart rate increase and blood flow is diverted to the heart and muscles, a response that, in evolutionary terms, was only ever meant to be triggered by physical danger. In short many people's blood pressure rises whenever they talk to a fellow human and the more difficult they find communication in general, the more precipitous the rise. In time this leads to physiological exhaustion and a biologically based need to withdraw from others for self-preservation, which ironically results only in loneliness and isolation, and ultimately premature disease and death, a condition Lynch has named the Physiology of Exclusion.

A similar rise in blood pressure occurs in children who have to read aloud, be it at school, at home, or on the stage, but, as Lynch, Katcher, and other colleagues discovered, just the simple presence of a pet canine lowers it. Further, Katcher found that gazing at fish swimming in a tank caused his volunteers' blood pressure to drop more than any traditional meditation techniques they practiced.

Pets and Our Health

New scientific research presented at the 10th International Conference on Human–Animal Interactions, in October 2004, revealed conclusively that pet-owners visit their doctors at least 10 per cent less than non-pet-owners.

Companionship and love may explain why pet-owners live longer.

The strongest evidence came from a German Socio-Economic Panel social science survey, which interviewed 10,000 people in 1996 and again in 2001. As most of the questions concerned income and work, the respondents could not possibly have guessed that the real aim was to evaluate the effects of pet ownership. Controlling results for sex, age, marital status, and so on, researchers found those who continuously owned a pet reported the least doctor visits, those who acquired a pet during the five-year period the next least, and those who had never had a pet or ceased to have a pet during the period, reported the most.

"I don't think you can give any single reason why pet-owners live longer, but I think companionship has a lot to do with it," said speaker Bruce Headey, from the University of Melbourne. Earlier he had written in the Medical Journal of Australia: "About 50 percent of adults and 70 percent of adolescents who own pets report that they confide in them. It is most unlikely that all this communication and companionship is wasted."

We know, although we may not always want to admit it, that our emotional and physical well-being is dependent upon our closeness to other species, a need put eloquently into words, and named the biophilia hypothesis, by the revered zoologist and environmentalist Edward O. Wilson: "For human survival and mental health and fulfilment we need the natural setting in which the human mind almost certainly evolved and in which culture has developed over these millions of years of evolution."

125

Most of us urban, suburban, and even rural dwellers are far from our natural setting. Utterly divorced from nature, we cannot see the stars for light pollution; living creatures are ruthlessly excluded—many, such as hedgehogs and bats, because the places they roost or hibernate in have been tidied out of existence to make way for manicured lawns and perfectly maintained real estate, while others such as pigeons and moles are persecuted, poisoned, and maimed because they "make a mess." So far from our natural setting, can it be a surprise that in the UK alone one in four people seek help for mental health problems at some point during their lives?

One way humans return, at least partially, to this closeness is by using their pets as a bridge to nature, by including them in human celebrations. One perfect case of this is the Cat Mitzvah of thirteen-month-old feline Fifi Katz attended by 90 humans, and celebrated in far away Israel by a tree being planted in her honor.

Some living creatures are excluded from the modern world
by the destruction of their roosting and hibernation sites.

There was fun and laughter at the Cat Mitzvah, which included parodies of Fiddler on the Roof and delicious food for all—but there was no mistaking that the true purpose of the occasion, as Fifi's "father" explained, was to honor "the beauty of creation as manifested in a particular little animal and to realize our own 'at-homeness' in the universe."

Cats seem instinctively to recognize our need and spontaneously use their physical beings to heal us emotionally and physically. One friend told me that when she and her long-term lover split, leaving her distressed and bereft, her cat, who had never before showed the slightest inclination to share her bed, curled up with her every night for several weeks. When the feline correctly judged her human was feeling better, she returned to her usual sleeping haunts, and has never returned to her human companion's bed.

Cats recognize our emotional needs, comforting us in times of distress.

Illustrator Csaba Pasztor's fluffy black feline Cica always knows not only when he is ill, but which area of his body is afflicted. Should his throat be sore she places her soft, warm paws on it; should his stomach ache she lies there or sometimes stands and uses her paws in a paddling motion to gently massage him. Pasztor finds this extraordinarily comforting and says Cica warms and calms him. When Pasztor first cohabited with Cica, he was allergic to cats but he loved her so much he determined to overcome this bar to feline friendship. Eschewing antihistamines or other scientific means, Pasztor buried his face in her fur, breathed in the perfume of her flesh, the scent of her coat, and, to use his own words, he "inoculated" himself. While this approach is not medically advisable, Csaba has never been allergic to cats since.

How Cats Help Older People

Cats are increasingly used to cheer up the elderly and frail, their furry, tactile bodies affirming continued connection with life through the unconditional affection that these perceptive creatures give. Psychologically, they can induce elation and give the housebound a sense of purpose, particularly if they are allowed to help look after their feline friend. Studies in Italy show that interacting with a feline for just an hour, three times a week for six weeks, improved nursing home residents' depressive symptoms as well as decreasing their blood pressure. A more anecdotal survey carried out by the UK charity Cats' Protection revealed that cats successfully helped the elderly overcome stress and loneliness, and that a staggering 75 per cent preferred the company of cats to that of other human beings.

Animal therapy

The first recorded use of animal-assisted therapy comes in 1792 from The Society of Friends, a Quaker Group at their York Retreat in England. Prevailing conditions in mental institutions were then beyond inhumane; the occupants were degraded, isolated, and all too often reduced to screaming hysteria. William Tuke responded by giving his patients positive reinforcement for self-control and making every effort to normalize the environment of these unfortunates—which included allowing them to wear their own clothes instead of institutionalizing uniforms and encouraging them to work in his retreat's gardens, and look after its small animal inhabitants, such as rabbits and poultry.

By the middle of the twentieth century the psychologist Levinson was taking things much further, using pets as a kind of transitional relationship for disturbed children. Having communicated trustingly with a pet in their therapy situation, the children then find it much easier to establish some kind of real relationship with their therapist and finally with the outside world. As might be expected, it was "the young nonverbal child, the inhibited, the autistic, the withdrawn, the obsessive–compulsive, and culturally disadvantaged" who responded most positively.

Following in Levinson's footsteps, Samuel and Elizabeth Corson and Leo Bustad began to use pets as therapists for troubled adults, such as Marie who "had no family and friends, would not communicate, and remained curled in the fetal position with no interest in living. She also had sores on her legs from continual scratching. When other measures failed, she was moved in with Handsome [the resident cat]. Whenever she began to scratch her legs, the cat played with her hands and distracted her. Within a month the sores were healed. She began to watch the cat and talk to the staff about him. Gradually she invited other residents in to visit with him. Now she converses with strangers as well as the nursing-home staff, about the cat and other subjects."

As Beck and Katcher point out in *Between Pets and People*, most people who are damaged have been hurt by words. The cat does not criticize but instead gives safety, refuge, acceptance, and an unambiguous love and, by having no words, wonderfully can "call forth speech from those who have given up speaking."

The Delta Society, a US charity, has established a Pet Partner's Program that gives animals and their humans the skills they need to visit people in need in myriad situations. Some of the tests felines must pass before commencing their healing career are; to be happy being carried in their handler's arms or a basket; to accept friendly petting from strangers; to stay on any lap they are placed; to exhibit calm when carried through busy areas or when faced with distracting circumstances.

Eight-year-old Moochie, a rising Delta star, belongs to Williamson, a nurse who, when she went part-time, decided to join Pet Partners. "I really feel it's kind of a spiritual thing, that closeness, that bonding that you get," she says of Moochie. "Previously I had dogs which I took to nursing homes, but I never knew a cat could do this. What's impressive about Moochie is that he seems to know when people are sick. He tends to be very close to them. At a health facility, he knows he's going to do a special job. At home he's as wild as a March hare."

Moochie arrived seemingly too late for one terminally-sick patient who had already drifted into a coma, but nonetheless Williamson put him "on the bed and rubbed him real hard by the man's face. He was just lying there with his arms under the sheets. As he felt Moochie, the man awoke from the coma, took his arms out from under the sheet and began petting him. The man's wife and daughter were elated at the result."

The Delta Society has established a Pet Partner's Program that trains animals and their humans in the skills of healing.

But it is not just the cat's presence that helps us—so does studying her genetics. Natural selection and adaptation to her wonderful wild environment has made her robust and able to resist the serious diseases such as cancers, AIDS, and multiple sclerosis that affect us. This means the sequencing her genetic map, which is being carried out by a team headed up by Stephen O'Brien at the National Cancer Institute in Maryland, is proving simply invaluable for cat and human alike.

Through millions of years of mammalian evolution certain genetic sequences have been conserved both in humans and felids. This probably means they are extremely important for our health and survival and allows scientists to target them for what may well prove to be breakthrough research to improve our well-being.

The gene for black fur is originally a mutation; and a mutation that is evidently a successful accident.

But it isn't just conserved gene sequences that are interesting to science—so are genetic mutations. O'Brien, who has been studying feline genetics for decades, was interested in the fact that most wild cats have a genetic phenotype—visual characteristic—connected with their coloration and pattern of their coat in common with the domestic cat. The striped tabby matches the tiger, while the blotched tabby resembles the clouded leopard. He wanted to know if these mutations had occurred once in feline evolution and simply remained, or if they occurred spontaneously at different times. O'Brien decided to study three species of felines, some but not all of whose members possess black coats—*Felis catus*, the jaguar, and the South American jaguarundi—and discovered different gene variants for black fur in each lineage. This means the mutation must have appeared repeatedly, not once, and so surely have some distinct evolutionary advantage, particularly as it is a recessive gene—meaning both mother and father must possess it for the quality to appear in their offspring.

But what was the black gene for? Could it be for camouflage? Certainly the black feral cats who now haunt the shadow lands of big cities might be at an advantage compared to their tabby relatives but in the forest and jungle broken patterning is a far more effective disguise for an ambush predator. Perhaps then, O'Brien speculated, the gene confers some other important benefits such as overall health or resistance to disease? And indeed the mutation is proven to affect a protein gene, which viruses use to get inside cells and cause disease, and which, interestingly, is from the same family as one which enables resistance to HIV in humans. "There is a mutation in humans that knocks this gene out and causes complete resistance to HIV," O'Brien revealed in an interview with Kristen Philipkoski:. He went on, "So it may be that these cats have a high frequency of black because they [achieved] some sort of resistance by blocking some infectious agent." O'Brien hopes that by deepening our understanding of the natural defences felines use against disease we will be able to gain insights into treating those same complaints in us.

Cats and Forensic Science

Unraveling the cat genome has had unexpected repercussions for the capture of criminals. On October 3 1994, Shirley Dugay disappeared from her home in Prince Edward Island, Canada. Not long afterward her car, complete with blood stains matching those of the thirty-two-year old mother, was found abandoned. Three weeks later a man's leather jacket covered in the victim's blood and 27 cat hairs was discovered in the woods nearby. Then in May 1995 Dugay's body was discovered in a shallow grave. The chief suspect in the gruesome crime, Dugay's estranged common-law husband Douglas Beamish, was arrested. But how could the Canadian Mounties link him conclusively to to the crime? It was then that one officer remembered that Beamish lived with a cat, a white American shorthair named Snowball.

The laboratory of Genomic Diversity in Frederick, Maryland, where Stephen O'Brien and others were unraveling feline genetics, was asked to determine whether genomic DNA from the cat hairs in the jacket matched those belonging to Snowball. Marilyn A. Menotti-Raymond typed in the DNA from 27 hairs selected for optimal performance in forensic analysis. They were a perfect match for those in Snowball's blood. Analysis of genetic material from 19 other feline residents of Prince Edward Island, and nine cats from around the United States, proved sufficiently different for the laboratory to be certain the hairs on the jacket could have come only from Snowball.

Beamish was convicted of second-degree murder on July 19 1996, and a legal precedent was set.

GLOSSARY

POPULAR AND TRADITIONAL CAT BREEDS

The Moggie

Most of the cats who strut with velvet paws across unforgiving concrete and lush meadow alike are moggies—cats of no determined breed. They may be tortoiseshell, silver, or black; tiny or great; shy or outgoing, but every single one is beautiful. They possess an unparalleled genetic diversity giving them health, bounce, vigor, and intelligence. The truly essential cat: paradoxical, infuriating, and charming by turn, who could resist her?

THE LONGHAIRS

European travelers were immediately captivated by the gorgeous longhairs of the East and brought them back to grace the courts of Europe.

Angoras

(See pages 37–39 for details on the modern-day "Angoras" and Turkish cats, which are somewhat different from this founding breed.)
Tall, rangy, and sinuous, these cats originate in Ankara, Turkey. In their homeland these felines come in colors ranging from red-tabby through white. Their fur is long, silky, and sleek but unlike the Persian it lacks an undercoat. Their popularity in the West has been eclipsed by the fuller-coated Persian.

Persians

These exotic and glamorous creatures were first discovered in Ishfahan, Persia—modern-day Iran. Distinctly Oriental in demeanor, with fetching Fu Manchu-type whiskers, they come in a wide range of strikingly mixed colors, such as a pale blue-gray coat mixed with deep orange eyes (reputedly a favorite with Queen Victoria)white coat with pale china blue eyes, and jet black with copper eyes. Their extremely profuse coat, which in contemporary times resembles a puff-ball, needs grooming for at least fifteen minutes a day. Normally phlegmatic in temperament, Persians are ideal for those who like a quiet life.

Chinchilla

In the UK all longhairs are classified as separate breeds, while in America the delicate Chinchilla is classified within the Persian group. Eye-lids outlined in black emphasize deep green irises, bringing to mind the striking eyes of the Egyptian cats seen in their ancient art.

Himalayans (Longhair Colorpoints)

A genetically engineered breed, Himalayans originate from crosses between the stunning Siamese and smoke, silver, tabby, and black Persians. Five years of selective breeding between these offspring gave rise to Debutante, a long-haired with contrasting colored mask, ears, tail and paws. Over time further selective breeding gave the Himalayans the characteristic stocky body of the Persian.

Birmans

Legend has it that 100 white cats and an aged priest called Mun-ha lived in the ancient Burmese temple of Lao-Tsun. One night Thai raiders attacked the temple killing the priest. His own pet, Sinh, jumped onto his body and faced the statue of the temple's golden-hued, blue-eyed goddess Tsun-Kyan-Kse. Being truly enlightened, the priest's soul entered the cat whose coat immediately took on the same blondish color as the goddess's body and whose eyes became as sapphires. Except where Sinh's paws touched his beloved master's body his legs became brown. By morning all the temple cats were transformed and, as guardians of priestly souls, were elevated to the sacred.

Now bred in several colors, they are distinguished from Himalayans by the elegant white "gloves" on their paws.

Balinese

Named after Balinese dancers because of their graceful carriage, in reality this breed originated in the US and is simply a longhaired Siamese. It was developed by breeding together the longhaired kittens who appear in Siamese litters from time to time. As longhair is a recessive gene this means that Balinese always breed true as

they do not possess the dominant gene for short hair.

Maine Coon

These tough sturdy mousers (see page 43) originated in Maine, USA, where they were, and remain, a firm favorite with farmers. The appellation "Coon" was added because their magnificent bushy tail resembles that of the raccoon, a populous local creature. They possess attractive ear tufts and are most popular in their rich, brown, tabby guise.

Norwegian Forest Cat

Very similar to the Maine Coon, this cat evolved hundreds, perhaps thousands, of years ago in the hard, cold, wet conditions of Norway's woodlands and is even featured in Norse mythology and folklore as a fairy cat. She possesses a warm undercoat and a gleaming top coat, which makes her relatively comfortable even in driving rain—something the majority of felines abhor. Unlike the Persian she requires no grooming.

THE SHORTHAIRS

The Exotic Shorthair

A Persian by any other name, these cats come in all the glorious color combinations of her longhaired relative. They also possess the stocky and gorgeously cuddly body shape and relaxed temperament of the Persian.

British, European, and American Shorthairs

Legend has it that the Roman Legions brought cats to England 2,000 years ago (see page 31). Admired for their independence and valued as ratters, they were later taken to America by sailors, indeed some speculate that the Mayflower itself carried a feline cargo. Heavy in build and with dense coats, ranging from silver tabby through pale cream to dark brown, devotees say that champion specimens are far superior to their moggie counterparts. Devotees of moggies may well demur.

Abyssinians

This enigmatic looking cat, similar in aspect to some of the wild cats of northern Africa, has a unique coat. Thanks to a mutant gene that endows every one of her delicious dark brown hairs with two or more dark bands, she has a speckled, ticked appearance very similar to that of a wild rabbit. Before the name Abyssinian was fixed upon she was also known as Bunny Cat and Hare Cat. The Somali is a longhaired Abyssinian, whose hairs may be encircled by up to twenty black bands.

The Egyptian Mau

A natural spotted tabby (most domestic cats with spotted coats are the result of selective breeding), this cat's coat is exquisite as befits one who is probably a direct descendent of the African wildcat (see page 30) and was revered in ancient Egypt.

Russian Blue

Legend has it that this blue cat with an upstanding coat and large upright ears originated from the romantically named port of Archangel on the Arctic Ocean, from whence she was taken to western Europe by sailors. In the 1940's and '50's British cat fanciers bred her with blue point Siameses to give her a more oriental body type. They were so successful that the breed standards had to be rewritten. However, in the 1960's they were bred selectively once more to regain their original, chunkier body type.

Korat

With its unique and appealing heart-shaped face and lively, large green eyes, the Korat still resembles its ancestors from its native Thailand. Her coat is colored a deep, steely gray but light reflected from the fine, silvery tips of every hair appears to give her a magical halo.

Burmese

Burmese cats are virtually all descended from Wong Mau, a genuine Burmese brought from Rangoon, who was mated with a Seal point Siamese. The offspring were then mated with their mother and one another. The kittens were variable in color but sufficiently resembled Wong

Mau, having a dark brown body with darker points, for a separate Burmese breed to be established during the 1930's. Various further pairings have now lead to a variety of colors in this extremely sociable cat. For those who favor a dog-like cat, Burmese could be ideal as they love to retrieve!

Siamese

In their native land the majority of Siamese (see pages 33–37) have kinked tails, which, according to myth, were used for carrying their human's golden rings, but in the west, cat fanciers have more or less eradicated this rather fetching characteristic. Striking in the extreme, their colors range from cream with seal points to snowy white with pastel pinkish-gray points, known as lilac.

Siamese, like Burmese, are very sociable cats who really enjoy talking—making them ideal language students.

The longhaired version of the Siamese is the Balinese.

BIBLIOGRAPHICAL REFERENCE

The following books, articles, and journals provided invaluable reference for the writing of this book. I am greatly indebted to their authors and I hope that you might, in turn, look up their work. (Each chapter is given its own complete listings and therefore reference duplication between chapters is intentional.)

CHAPTER ONE

Anton & Turner *The Big Cats and Their Fossil Relatives* (Columbia, 1997)

Bökönyi, S. *History of Domestic Mammals in Central and Eastern Europe* (Budapest, 1974)

Clutton-Brock, Juliet *British Museum Book of Cats* (London, 1994)

Collier G.E. & O'Brien, S.J. "A Molecular Phylologeny of the Felidae Immunological Distance" *Evolution* (Volume 39, 1985, pp.473–487)

Conway, W.M. "The Cats of Ancient Egypt" *English Illustrated Magazine* (Volume 7) *Encyclopedia Sinica*

Hemmer *Carnivore* (Volume 1, 1997)

Heroditus *The Histories*

Janssen, Rosalind and Jack *Egyptian Household Animals* (London, 1989)

Johnson W.E. & O'Brien S.J. "Phylogenetic Reconstruction of the Felidae" *Journal of Molecular Evolution* (Volume 44 [supp1], 1997, pp.S98–S116)

Kircher, A. *China Illustrata*, translated by Dr. Charles D. van Tuyl (Muskogee, Oklahoma, 1987)

Malek, Jaromir *The Cat in Ancient Egypt* (British Museum Press, London, 1993)

O'Brien, S.J. et al *The Feline Genome Project,* Lab of Genomic Diversity, Maryland (2002)

O'Brien et al "Setting the molecular clock in the Felidae," *Tigers of the World*, 4th edition, RL Tilson & US Seal (Noyes, New Jersey, pp.10–27)

Randi, E. & Ragni, B. "Genetic Variability & Biochemical Systematics of Domestic & Wildcat Populations" *Journal of Mammalogy* (Volume 72, 1991, pp. 79–88)

Saunders, Nicholas J. *Cult of the Cat* (Thames and Hudson, London, 1991)

Smithers, Reay "Cat of the Pharaohs" *Animal Kingdom* (Volume 71, 1968)

Sunquist & Sunquist *Wild Cats of the World* (Chicago, 2002)

Tabor, Roger *The Wild Life of the Domestic Cat* (London, 1984)

Van Vechten *A Tiger in the House* (London, 1921)

de Visser, Dr M.W. "Dog & Cat in Japanese Folklore" *Transactions of the Asiatic Society of Japan* (Volume 37, 1909)

Volker, T. *The Animal in Far Eastern Art* (Leiden, 1950)

Weir *Our Cats* (London, 1889)

Wilson Quarterly archive

CHAPTER TWO

Barthez, P.Y. et al "Prevalance of Polycystic Kidney Disease in Persain and Persian Related Cats in France" *Journal of Feline Medice and Surgery* (Volume 5, 2003, pp.345–7)

Bingley *Animal Biography* (London, 1813)

Birkenmeier, E. & Birkenmeier, E. "Hand Rearing the Leopard Cat" *International Zoo Yearbook* (Volume 11, 1971, pp.118–121)

Bökönyi, S. "Archaeological Problems and Methods of Recognising Animal Domestication" *The Domestication and Exploitation of Plants and Animals*, Ucko, P.J. & Dimbleby, G.W. (ed.) (London, 1969)

Burger I (ed.) *Pets, Benefits and Practice* London British Veterinary Association Publications (pp.25–30)

Clutterbuck, Martin R. *The Legend of Siamese Cats* (White Lotus Press, Bangkok, 1998)

Collier, V.W.F. *Dogs of China and Japan in Nature and Art* (London, 1921)

Critchley, Thomas *Persian Cat* (Interpet, London, 2002)

Feathestone, Heidi J. and Sansom, Jane "Feline corneal sequestra, a review of 64 cases from 1993-2000" *Veterinary Opthalmology* (Volume 7, 2004, pp.213–227)

Fogle, Bruce *The Cat's Mind*, (Pelham Books, London, 1991)

Mivart *The Cat* (London, 1881)

Narfström, K. "Hereditary and Congenital Ocular Disease in the Cat Review" *Journal of Feline Medice and Surgery* (Volume 1, 1999, pp.135–41)

Paradise de Moncrif, Francois-Augustin *Histoire des Chats* (Paris, 1727)

Randi, E. & Ragni, B. "Genetic Variability & Biochemical Systematics of Domestic & Wildcat Populations" *Journal of Mammalogy* (Volume 72, 1991, pp.79–88)

Simpson, Frances *The Book of the Cat* (London, 1902)

Turner, Dennis C. "Human Cat Interactions: Relationships with, and Breed Differences between, Non-pedigree and Persian & Siamese Cats" *Companion Animals and Us*, Podberscek, Paul & Serpell (ed.) (Cambridge, 2000)

Turner, Denis C. and Bateson, Patrick (ed.) *The Domestic Cat: The Biology of its Behaviour* (Cambridge, 2000)

Weir *Our Cats* (London, 1889)

Zeuner, F.E. *History of Domesticated Mammals* (London, 1963)

CHAPTER THREE

Cameron-Beaumont *Visual Tactile Communication in the Domestic Cat and Undomesticated Small Felids* (University of Southampton, 1997)

Estes, R. *The Safari Companion: A Guide to Watching African Mammals* (Chelsea Green Publishing Company, Vermont, 1993)

Greene, David *Incredible Cats* (London, 1984)

Harrington, E. "Panthera leo," *Animal Diversity* website (2004)

Leyhausen, Paul *Cat Behavior, the Predatory and Social Behavior of Domestic and Wild Cats* Taylor and Francis (1978)

Masson, Jeffrey *The Nine Emotional Lives of Cats* Vintage (London, 2003)

Mellon, I.M. *The Science and Mystery of the Cat* (1940)

Schaller, G. *The Serengeti Lion* (The University of Chicago Press, Chicago, 1972)

Tabor, Roger *The Wild Life of the Domestic Cat* (London, 1984)

Turner, Denis C. and Bateson, Patrick (ed.) *The Domestic Cat: The Biology of its Behaviour* (Cambridge, 2000)

Tabor, Roger *The Wild Life of the Domestic Cat* (London, 1984)

Turner, Den is C. and Bateson, Patrick (ed.) *The Domestic Cat: The Biology of its Behaviour* (Cambridge, 2000)

Van Vechten *A Tiger in the House* (London, 1921)

Wright, Michael and Walters, Sally (ed.) *The Book of the Cat* (Summit Books, New York, 1980)

CHAPTER FOUR

Cameron-Beaumont *Visual Tactile Communication in the Domestic Cat and Undomesticated Small Felids* (University of Southampton, 1997)

Caro, T.M. "The Effects of Experience on the Predatory Behaviour of Cats" *Behavioural and Neural Biology* (Volume 29, 1980, pp.1–28)

Collier, G. et al "Food Optimisation of Time and Energy Constraints in the Feeding Behavior of Cats, A Lab Simulation" *Carnivore* (Volume 1, part 1, January 1978)

Collier, G. and Johnson D.F. "Meal Patterns of Cats Encountering Variable Food Procurement Costs" *Journal of the Experimental Analysis of Behaviour* (Volume 67, 1997, pp.303–10)

Crowell-Davis, S. et al "Social Behavior and Aggressive Problems of Cats" *Veterinary Clinics of North America: Small Animal Practice* (Volume 27 No. 3, May 1997)

Masson, Jeffrey *The Nine Emotional Lives of Cats* (Vintage, London, 2003)

CHAPTER FIVE

Bechterev, W. "Direct Influence of a Person upon the Behaviour of Animals" *Journal of Parapsychology* (Volume 13, 1949)

Clutterbuck, Martin R. *The Legend of Siamese Cats* (White Lotus Press, Bangkok, 1998)

Domhoff, G. William "Using Content Analysis to Study Dreams: Applications and Implications for the Humanities," *Dreams*, Kelly Bulkeley (ed.) (Palgrave, London, 2001)

Eason, Cassandra *The Psychic Power of Animals* (London, 2003)

Fitzpatrick, Sonya *What the Animals Tell Me* (London, 1999)

Greene, David *Incredible Cats* (London, 1984)

Janssen, Rosalind and Jack *Egyptian Household Animals* (London, 1989)

Katcher, A.H. and Beck, A.M. (ed.) *New Perspectives on Our Lives with Companion Animals* (University of Pennsylvania Press, Philadelphia, 1983)

O'Donnell Elliot *Animal Ghosts*

Peretz, Isabelle and Zatorre, Robert (ed.) *The Cognitive Neuroscience of Music* (Oxford University Press, Oxford, 2004)

Schneider, A. & Domhoff, G. W. *The Quantitative Study of Dreams* www.dreamresearch.net (2004)

Sheldrake, Rupert *Dogs Who Know When Their Owners Are Coming Home* (Crown, London, 2003)

Sheldrake, Rupert *The Sense of Being Stared At* (Crown, London, 2003)

Shou, Zhonghao "The Haicheng Earthquake and Its Prediction" *Science and Utopya* (Volume 65, 1999)

Smythe, R.H. *How Animals Talk* (London, 1959)

Tabor, Roger *The Wild Life of the Domestic Cat* (London, 1984)

Van Vechten *A Tiger in the House* (London, 1921)

de Visser, Dr M.W. "Dog & Cat in Japanese Folklore" *Transactions of the Asiatic Society of Japan* (Volume 37, 1909)

Weir *Our Cats* (London, 1889)

Winer, Gerald A. and Cottrell, Jane E. "Does Anything Leave the Eye when we See? Extramission Beliefs of Children and Adults" *Current Directions in Psychological Science* (Volume 5, 1996, pp.137–42)

Winer, Gerald A. et al "Fundamentally Misunderstanding Visual Perception" *American Psychologist* (Volume 57, 2002, pp.417-24)

Wright, Michael and Walters, Sally (ed.) *The Book of the Cat* (Summit Books, New York, 1980)

Wylder, J.E. *Psychic Pets: the Secret World of Animals* (London, 1978)

CHAPTER SIX

Ballner, Maryjean *Cat Massage* (New York, 1997)

Beck, A.M. and Katcher, A.H. *Between Pets and People* Purdue (Indiana, 1996)

Headey, Bruce "Pet Ownership: Good for Health?" *Medical Journal of Australia* (Volume 179, No. 9 pp.460–1)

Katcher, A.H. and Beck, A.M. (ed.) *New Perspectives on Our Lives with Companion Animals* (University of Pennsylvania Press, Philadelphia, 1983)

Levinson, B. *Pets and Human Development* (Springfield, Ill. ,1972)

Podberscek, E.S. et al (ed.) "Personality Research on Pets and Their Owners" in *Companion Animals and Us* edited by A.L. Podberscek, E.S. Paul, and J.A. Serpell, (Cambridge University Press, Cambridge, 2000)

Serpell, James *In the Company of Animals* (Blackwell, London, 1996)

Soennichsen, S. and Chamove, A. "Responses of Cats to Petting by Humans" *Anthroxoös* (Volume 15, No. 3, 2002)

Stasi MF et al *Pet-therapy: a Trial for Institutionalized Frail Elderly Patients.*

Every effort has been made to list accurate data for the entries in this reference. The Publishers and the Author apologise for any inaccuracies or omissions, all of which are unintentional. We shall, if informed, correct future editions of this bibliography.

ACKNOWLEDGMENTS

I would like to thank the patient staff of Science Two South in the British Library and the wonderful London Library; my even more patient publisher Cindy Richards and her calm cohort Georgina; Alison Taylor, chief Behaviorist at Battersea Dogs Home; Stefania Baratti, a Battersea cat expert; Linda Tellington-Jones for her input on Tellington TTouch; my friends Sylvia Stickles and Csaba Pasztor for their insightful views of feline life; and all my other friends, too numerous to mention, who have contributed to this book more than they might ever realize.

INDEX

S

Sand cat (*Felis margarita*)
10, 13–14, 66
scent
glands 57, 58, 116
sensing 95–6
Schaller, George 66
Scottish Fold cat 44
scratching, and posts 57–8
Sekhemet, Egyptian goddess 24
selective breeding
coat mutations 43–5
dangers of 42–5
self-awareness, mirror test for
74–5, 76
sensuousness 81
Sheldrake, Rupert 98–9, 106,
109, 117
ships, cats and 91, 92
Siamese cat 33–5, 36–7, 41, 135
Siamese Cat Treatises 34–5, 100
silkworms 16, 18
Simpson, Frances, *The Book of
the Cat* 40
skin, sensitivity of 117
smell 95–6
use of 57–8
Smithers, Reay 21
Smythe, R.H. 93, 105
snakes 22–3
Society of Friends (Quakers)
128
Soennichsen, Susan 116
Somali cat 134
Sphinx cat (hairless) 44

spotted cats (Egyptian Mau)
36, 134
spraying, indoors 64–5
staring 35, 59, 109
static electricity 100
stroking 115, 116, 117
submission, apparent 62
Sumatra 98
Switzerland, Zurich University
research 48
symbolism 110–11

T

tabby cat
blotched 19, 33
striped 32–3, 34
Tabor, Roger 19, 33, 67,
77, 96–7
Tail Up, greeting 66–7, 68
tails, to show expressions
62, 68–9
teeth, conical 9, 10
telepathy, power of 105–8
Tellington Touch (T Touch)
120–1
Tellington-Jones, Linda 120
temperature 61
territory 83
home range 84–5
marking 64–5
Thailand, Siamese cats 33–5
Thebes, cat cemetery 26–7
thinking, cats' 72–3
tigers 11, 114–15
touch 112, 115, 123–4

trust 62
Tuke, William 128
Turkey, Angoran cats from 37–9
Turner, Dennis 48, 49

V

valerian 57–8
Vechten, Van 88
Vikings 92
vision 90–1

W

weather predictions 94–9
Weigold, Dr Hugo 15
Weir, Harrison 31–2, 33, 40–1
White, Gilbert 100
wild cats 12–17
Wilson, Edward O. 125
witches, cats and 101–2